SOMATIC CELL HYBRIDIZATION

SOMATIC CELL HYBRIDIZATION

Edited by

Richard L. Davidson
*Children's Hospital Medical Center
and Harvard Medical School,
Boston, Massachusetts*

and

Felix F. de la Cruz
*National Institute of Child Health and Human Development
Bethesda, Maryland*

A Monograph of the National Institute of Child Health and Human Development

Raven Press ■ New York

Made in the United States of America.

International Standard Book Number 0-911216-75-8
Library of Congress Catalog Card Number 74-75725

Proceedings of a Conference Sponsored by

The National Institute of Child Health and Human Development
National Institutes of Health
Bethesda, Maryland

March 11–14, 1973
Winter Park, Florida

Organizing Committee:

Richard L. Davidson, Ph.D., *Chairman*
Felix F. de la Cruz, M.D.
Park S. Gerald, M.D.
Hilary Koprowski, M.D.
John Littlefield, M.D.
Frank Ruddle, Ph.D.

Foreword

The occurrence of hybridization between somatic mammalian cells *in vitro* was first observed in 1960. Since that time, the techniques of somatic cell hybridization have been used in a rapidly expanding range of experiments, in areas as diverse as formal genetics and membrane dynamics. Because of the increasing diversification in the applications of the techniques of somatic cell hybridization, and because of the rapid pace of developments in many of the areas being studied with these techniques, it was decided to hold an international conference at which all the major areas of research involving cell hybridization could be considered together. The conference was sponsored by the National Institute of Child Health and Human Development and took place March 11–14, 1973, in Winter Park, Florida.

The conference was organized around four major areas of research: (1) gene mapping in hybrid cells, (2) viruses, cell membranes, and malignancy in fused cells, (3) gene regulation in hybrid cells, and (4) gene expression in heterokaryons. An extensive review of the published literature in each of these areas was presented, followed by workshop sessions in which recent results were described. The complete proceedings of the conference, including the discussions that followed each paper, are presented in this volume.

I would like to express my appreciation to the members of the organizing committee—Drs. Felix de la Cruz, Park Gerald, Hilary Koprowski, John Littlefield, and Frank Ruddle—for their help in planning the conference. I would especially like to acknowledge the contribution made by Dr. de la Cruz of the National Institute of Child Health and Human Development in arranging the conference and the publication of the proceedings. Finally, I would like to thank the editorial assistants—Drs. Gail Bruns and Diane Horn and my wife Jalane—who contributed significantly in editing the proceedings of the conference and preparing the final volume.

Richard L. Davidson, Ph.D.
Genetics Division
Children's Hospital Medical Center and
Department of Microbiology and
Molecular Genetics
Harvard Medical School
Boston, Massachusetts 02115

Preface

Somatic Cell Hybridization

The development of techniques for *in vitro* cultivation and fusion of somatic cells has provided a valuable tool for the study of mammalian genetics, including the genetics of man. Studies with cell hybrids are helping to construct the genetic map of man and are contributing to our knowledge of many of the basic processes carried out by mammalian cells, such as the regulation of gene expression and the control of cell growth.

Experiments with hybrid cells are providing information concerning the mechanisms of gene regulation in mammalian cells and are establishing systems that will be useful for studying, at the cellular level, the nature of the genetic defects leading to congenital malformations. The techniques presently being developed for the analysis of gene regulation in differentiated cells may allow one to determine precisely the genetic defects resulting in abnormal patterns of development and may thereby lead to ways of preventing these abnormalities.

Studies with cell hybrids have provided investigators with an important method of studying oncogenesis and in understanding the relationship between viruses and cancer. Experiments have already shown that complete viral genomes are maintained and can be reactivated in cells transformed to the malignant state by viral treatment. Experiments are also in progress to determine where in the human genome tumor viruses are attached. Other hybridization experiments are analyzing the genetic changes that occur when a normal cell becomes a cancer cell.

The mapping of the human genome, that is, the determination of the chromosomal location of specific genes, is proceeding rapidly using cell hybridization techniques. These techniques have the advantage over mapping by pedigree analysis in that information can be obtained by studying events that occur over a period of weeks rather than decades. Furthermore, classical techniques for gene mapping preclude controlled matings that could provide the genetic information sought by the investigator. In addition to its importance for basic studies on human genetics, knowledge of the human genetic map could be used for the early diagnosis of genetically determined diseases in which the clinical manifestations of the condition are not apparent until later in life. For example, if the genes for Huntington's chorea could be linked to other genes whose activities could be monitored in cell culture, it might be possible to determine with amniotic fluid samples (especially in high-risk families) whether a particular individual will suffer from the condition. Such determinations could be made by linkage inference at a time when it would not be possible to analyze directly for the disease. Knowledge of the human genetic map could also result in a better understanding of the genetic defects resulting from chromosomal abnormalities such as translocation and nondisjunction.

Realizing the importance of somatic cell hybridization in our understanding of various human disorders—particularly mental retardation, degenerative diseases involving the central nervous system, teratology, and cancer—the National Institute of Child Health and Human Development sponsored a conference in this important area. The conference was held in Winter Park, Florida, on March 11–14, 1973, and was attended by leading investigators in this field of research. This volume contains the papers that were presented and the significant discussions of the conference.

I would like to thank the conference planning committee for developing the program and for assembling such an outstanding group of participants. I particularly want to thank Dr. Richard L. Davidson, the Chairman of the Planning Committee, for all his efforts in organizing the conference and editing the conference proceedings.

Felix de la Cruz, M.D.
Special Assistant for Pediatrics
Mental Retardation Program
National Institute of Child Health and Human Development

Contents

Conference Participants

Dr. Giuseppe Attardi
Division of Biology
California Institute of Technology
Pasadena, California 91109

Dr. Claudio Basilico
Department of Pathology
New York University Medical Center
School of Medicine
New York, New York 10016

Dr. D. Bootsma
Department of Cell Biology and Genetics
P.O. Box 1738
Medical Faculty of Rotterdam
Rotterdam, The Netherlands

Dr. Teh-Sheng Chan
Department of Biology
Massachusetts Institute of Technology
Cambridge, Massachusetts 02139

Dr. Harvey Colten
Division of Immunology
The Children's Hospital Medical Center
300 Longwood Avenue
Boston, Massachusetts 02115

Dr. Hayden Coon
Laboratory of Cell Biology
National Cancer Institute
National Institutes of Health
Bethesda, Maryland 20014

Dr. Carlo Croce
The Wistar Institute
36th Street at Spruce
Philadelphia, Pennsylvania 19104

Dr. Gretchen Darlington
Department of Biology
Kline Biology Tower
Yale University
New Haven, Connecticut 06520

Dr. Richard L. Davidson
Genetics Division
The Children's Hospital Medical Center
300 Longwood Avenue
Boston, Massachusetts 02115

Dr. Felix F. de la Cruz
Mental Retardation Program
National Institute of Child Health and Human Development
National Institutes of Health
Bethesda, Maryland 20014

Dr. George N. Donnell
Division of Medical Genetics
Children's Hospital of Los Angeles
4650 Sunset Boulevard
Los Angeles, California 90054

Dr. Michael Edidin
Mergenthaler Laboratory for Biology
The Johns Hopkins University
Baltimore, Maryland 21218

Dr. Boris Ephrussi
Centre de Genetique Moleculaire
91. Gif-sur-Yvette
France

Dr. John Evans
MRC Clinical and Population Cytogenetics Unit
Western General Hospital
Creve Road
Edinburgh, EH4 2XU, Scotland

Dr. Marc Fellous
Genetics Laboratory
Department of Biochemistry
University of Oxford
Oxford, England

Dr. Jean Frezal
Hopital des Enfants Malades
149, rue de Sevres
Paris XV, France

Dr. H. Galjaard
Department of Cell Biology and Genetics
P.O. Box 1738
Medische Faculteit, Rotterdam
Rotterdam, The Netherlands

Dr. William J. Gartland
Research Grants Branch
National Institute of General Medical Sciences
National Institutes of Health
Bethesda, Maryland 20014

Dr. Thomas Gelehrter
Department of Human Genetics
School of Medicine
Yale University
New Haven, Connecticut 06520

Dr. Park S. Gerald
Genetics Division
The Children's Hospital Medical Center
300 Longwood Avenue
Boston, Massachusetts 02115

Dr. Saimon Gordon
Rockefeller University
New York, New York 10021

Dr. Christopher Graham
Department of Zoology
University of Oxford
South Parks Road
Oxford OX1 3PS, England

Dr. Morgan Harris
Department of Zoology
University of California
Berkeley, California 94720

Dr. David L. Joftes
Mental Retardation Program
National Institute of Child Health and Human Development
National Institutes of Health
Bethesda, Maryland 20014

Dr. Barbara Knowles
The Wistar Institute
36th Street at Spruce
Philadelphia, Pennsylvania 19104

Dr. Hilary Koprowski
The Wistar Institute
36th Street at Spruce
Philadelphia, Pennsylvania 19104

Dr. Robert Krooth
Department of Human Genetics and Development
College of Physicians and Surgeons
Columbia University
New York, New York 10032

Dr. John Littlefield
Genetics Unit
Massachusetts General Hospital
Boston, Massachusetts 02114

Dr. Henry R. Mahler
Department of Chemistry
Indiana University
Bloomington, Indiana 47401

Dr. Barbara Migeon
Department of Pediatrics
The Johns Hopkins University
School of Medicine
Baltimore, Maryland 21205

Dr. Orlando J. Miller
Department of Human Genetics and Development
College of Physicians and Surgeons
Columbia University
New York, New York 10032

Dr. John Minna
Laboratory of Biochemical Genetics
National Heart and Lung Institute
National Institutes of Health
Bethesda, Maryland 20014

Dr. G. Pontecorvo
Imperial Cancer Research Fund
Lincoln's Inn Fields
London WC2
England

Dr. Theodore Puck
Eleanor Roosevelt Institute for Cancer Research
University of Colorado Medical Center
Denver, Colorado 80220

Dr. James Renwick
London School of Hygiene and Tropical Medicine
Kepper Street
London WC1E
England

Dr. Nils Ringertz
Institute for Medical Cell Research and Genetics
Karolinska Institutet
S-104 01 Stockholm 60
Sweden

Dr. Frank Ruddle
Department of Biology
Kline Biology Tower
New Haven, Connecticut 06520

Dr. Thomas Shows
Roswell Park Memorial Institute
Buffalo, New York 14203

Dr. James B. Sidbury, Jr.
Department of Pediatrics
School of Medicine
Duke University
Durham, North Carolina 27710

Dr. Selma Silagi
Department of Obstetrics and Gynecology
New York Hospital—Cornell Medical Center
New York, New York 10021

Dr. Louis Siminovitch
Department of Medical Cell Biology
University of Toronto
Toronto, Ontario
Canada

Dr. Marcello Siniscalco
Department of Genetics
Faculty of Science
Via Mezzocannone 8
Naples, Italy

Dr. Armen Tashjian
Harvard School of Dental Medicine
188 Longwood Avenue
Boston, Massachusetts 02115

Dr. E. Brad Thompson
Laboratory of Biochemistry
National Cancer Institute
National Institutes of Health
Bethesda, Maryland 20014

Dr. Theodore D. Tjossem
Mental Retardation Program
National Institute of Child Health and Human Development
National Institutes of Health
Bethesda, Maryland 20014

Dr. John Watkins
Sir William Dunn School of Pathology
University of Oxford
South Parks Road
Oxford, England

Dr. Mary Weiss
Centre de Genetique Moleculaire
91. Gif-sur-Yvette
France

Dr. Francis Wiener
Department of Tumor Biology
Karolinska Institutet
S-104 01 Stockholm 60
Sweden

RAPPORTEURS:

Dr. Diane Horn
The Children's Hospital Medical Center
300 Longwood Avenue
Boston, Massachusetts 02115

Dr. Gail Bruns
The Children's Hospital Medical Center
300 Longwood Avenue
Boston, Massachusetts 02115

CONFERENCE STAFF:

Meryom Lebowitz
Conference Assistant
National Institute of Child Health and Human Development
National Institutes of Health
Bethesda, Maryland 20014

Ruth Markham
Secretary, Mental Retardation Program
National Institute of Child Health and Human Development
National Institutes of Health
Bethesda, Maryland 20014

Somatic Cell Hybridization, edited by
Richard L. Davidson and Felix de la Cruz.
Raven Press, New York © 1974.

Human Genetic Linkage and Gene Mapping by Somatic Cell Genetics

Frank H. Ruddle

Yale University, Kline Biology Tower, New Haven, Connecticut 06520

It has been suggested by a number of investigators that genetic information can be extracted from cultured somatic cells *in vitro* (28, 43, 61). These workers pointed out that somatic cells might undergo certain events that are basically parasexual in nature. Thus chromosome segregation in hybrids is analogous to meiotic segregation, and cell hybridization bears certain similarities to syngamy. The question may then be asked: can such analogies to sexuality be identified and then controlled for the purpose of genetic analysis in man and other mammalian species? The occurrence of parasexuality has been established in mammalian cell populations *in vitro*. Somatic cell hybridization, first demonstrated by Barski et al. (1), was a major early step in the formulation of somatic cell genetic systems, allowing as it does the combination of genetically different genomes within a single cell. Ephrussi and his co-workers (13) showed that hybrid combinations could be obtained between the cells of different species and that chromosomes of one or both parental genomes could be lost or segregated from the hybrid cell (14). Weiss and Green (69) first demonstrated the practical application of somatic fusion—segregation systems for the purpose of gene mapping in man. Other investigators have contributed useful procedures that enhance the formation of hybrid cells and their enrichment (22, 29). It is also possible to contemplate the use of additional parasexual mechanisms for somatic cell genetic analysis. Somatic (mitotic) recombination may occur in mammalian somatic cells (15, 57). Recent experiments suggest the possible occurrence of transformation (gene-transfer) mechanisms in mammalian cell cultures (31, 37, 40). Such systems might be employed ultimately for fine-resolution gene mapping in mammalian cell-culture systems.

One serious obstacle to genetic analysis in somatic cells has been the paucity of biochemical markers that are genetically determined and that are expressed in cultured somatic cells. This problem has been alleviated considerably during the past decade. There is now a relatively large catalog of biochemical genetic markers, which can be classified into the following broad categories: isozyme variants (51), salvage pathway, drug-resistant mutants (29), nutritional auxotrophs (7, 8, 24), temperature-sensitive mutants (36, 42, 63–65), recessive drug-resistant mutants (20), dominant drug-resistant mutants (21; Baker, *personal communication*) and surface antigens (27). At least a subset of these variants is under simple genetic control and can be obtained either by mutation selection *in vitro*, or as naturally occurring mutations or polymorphisms in populations of organisms.

Conditional lethal mutants are extremely useful in somatic genetic systems of analysis. Such mutants can be used to enrich for somatic hybrids (30, 48) as well as to limit the segregation of complementary genes and corresponding linkage units (49). In a system developed by Littlefield (29), one parental cell population lacks the enzyme thymidine kinase (TK) and, thus, is unable to utilize exogenous pyrimidine in the form of thymidine. The other parent is deficient for hypoxanthine-guanine phosphoribosyl transferase (HGPRT) and, thus, is unable to utilize exogenous purine in the form of hypoxanthine. Hybrids between these two lines are fully competent with regard to these functions because of complementation. If endogenous synthesis of pyrimidine and purine is blocked by application of aminopterin (A), the parental cells even in the presence of hypoxanthine (H) and thymidine (T) die, whereas the hybrids survive. Selection medium of this general type is termed HAT, and its use is based on the conditional, HAT-mediated killing of salvage pathway-deficient parental cells. Conditional lethals of the nutritional-auxotrophic (5, 24) and temperature-sensitive (63–65) types can be used in a comparable way.

Recently, dominant drug-resistant variants have been described (21; Baker, *personal communication*). These mutations can be combined with recessive conditional drug-resistant mutants. Such doubly mutated cells have special advantages for hybrid selection, since hybrids between such cells and wild-type cells can be recovered under nonpermissive conditions, whereas both parents are eliminated.

Rodent-human hybrids possess three properties essential to linkage analysis in man: the human chromosomes are preferentially segregated, the human chromosomes can be identified unambiguously, and human constitutive phenotypes are expressed and can be discriminated from their mouse counterparts. Various aspects of these properties have been recently reviewed in detail (48). Human-gene linkage can be accomplished by analyzing the segregation pattern of the human phenotypes and chromosomes in hybrid clones of independent origins. If phenotypes segregate together, this suggests that genes that control their expression are linked together on a particular chromosome. Linkage between genetic loci over the length of a chromosome has been termed *synteny*. A syntenic association does not specify a particular chromosome, nor does it indicate the map position or cytogenetic localization of loci. Loci can be *assigned* to particular chromosomes by correlating the segregation patterns of chromosomes and phenotypes. The assignment merely locates genes on specific chromosomes; it does not indicate position or map distance. One can make use of translocations or deletions in order to obtain gene positions on particular chromosomes. Regional mapping can make use of either chromosome rearrangements which occur in human parental cell populations or rearrangements which occur spontaneously in the hybrid cells (3, 45, 46). The above considerations have been considered in detail elsewhere (48, 49).

It has been possible to map 33 genes to 18 of the 24 human chromosomes by the methods outlined above. The existing human-linkage map limited to markers amenable to somatic cell genetic analysis is discussed in detail below. A similar, but less up-to-date listing has also recently appeared elsewhere (49).

Chromosome 1. PGM_1 – ENO – PEP-C – PGD

Van Cong et al. (66) have reported a syntenic relationship between phosphoglucomutase-1 (PGM_1) and peptidase C (PEP-C) using mouse-human hybrids. This synteny has been confirmed by Ruddle et al. (54). Westerveld and Meera Khan (70) have reported a synteny between 6-phosphogluconate dehydrogenase (PGD) and PGM_1 using Chinese hamster–human hybrids. Recently, Meera Kahn *(personal communication)* has reported a syntenic association between human 6PGD and enolase (ENO) in Chinese hamster–human hybrids. Taken together these findings imply that PGD, PEP-C, ENO, and PGM_1 are all syntenic. Ruddle et al. (54) have assigned PEP-C to chromosome 1 using mouse-human cell hybrids. This assignment has also been confirmed by the assignment of PGD to chromosome 1 independently by Bootsma *(personal communication)* and Hamerton et al. *(personal communication)* using Chinese hamster–human hybrids. If the findings based on cell hybrids are combined with linkages known from human-pedigree analysis, the following additional gene markers can be assigned to chromosome 1: zonular pulverulent cataract, Duffy blood group, auriculo-osteodysplasia, salivary amylase, pancreatic amylase, elliptocytosis, and Rhesus blood group. For complete literature citations see Ruddle et al. (54).

Chromosome 2. IDH – MOR

Preliminary results in our laboratory (10) suggest that isocitrate dehydrogenase (IDH) and NAD-malate dehydrogenase (MOR), which have been shown to be syntenic by Shows (58), can be assigned to chromosome 2. Since these loci disassociate at a relatively high frequency, they may possibly be distantly separated, perhaps on different arms.

Chromosome 3

No assignments.

Chromosomes 4 and 5. ADE-B

Kao and Puck (25) using Chinese hamster–human hybrids have demonstrated a positive association between hamster adenine B auxotrophy and a human B group chromosome when hybrids were propagated on minimal medium. The specific enzyme involved is unknown.

Chromosome 6. MOD – IPO-B – PGM_3

Chen et al. (6) using mouse-human hybrids have shown that cytoplasmic malate dehydrogenase (MOD) is assignable to chromosome 6. Tischfield et al. *(unpublished data)* have evidence to support a syntenic relationship between indophenol oxidase, tetrameric form B (IPO-B), and MOD. Recently, Westerveld *(personal communication)* has demonstrated an association between human PGM_3 and chromosome 6, using Chinese hamster–human hybrids.

Chromosome 7. MPI – PK_3

McMorris et al. (34) using mouse-human hybrids have assigned mannose phosphate isomerase (MPI) to chromosome 7. The correlation is not perfect since several clones have been detected that have chromosome 7 but lack apparent MPI activity. However, assignment can be ruled out for chromosomes other than 7, whereas a strong if not perfect correlation exists between 7 and MPI. Shows (60) has reported a syntenic relationship between MPI and the leucocytic form of pyruvate $kinase_3$ (PK_3).

Chromosomes 8 and 9

No assignments.

Chromosome 10. GOT

Creagan et al. (11) using mouse-human hybrids have evidence for the assignment of the cytoplasmic form of glutamate oxaloacetate transaminase (GOT) to chromosome 10.

Chromosome 11. LDH-A – EsA_4 – KA

Boone et al. (3) using mouse-human hybrids have assigned lactate dehydrogenase A (LDH-A) to chromosome 11. Shows (59) using mouse-human hybrids has reported a syntenic association between LDH-A and human esterase A_4 (EsA_4). Nabholz et al. (41) using mouse-human cells have reported a positive correlation between the segregation of LDH-A or -B activity and sensitivity of hybrid cells to antihuman cytotoxic antisera. Puck et al. (44) have reported on the segregation of a possibly similar human antigen(s) in Chinese hamster–human hybrid cells. They have reported it to be syntenic with LDH-A.

Chromosome 12. LDH-B – PEP-B – *m*-MOR – *m*-CS – GLY-A

Ruddle and Chen (55) and Chen et al. (6) using mouse-human hybrids have demonstrated a positive correlation between lactate dehydrogenase B (LDH-B) and chromosome 12. Hamerton et al. (19) have confirmed this assignment using Chinese hamster–human hybrids. Ruddle et al. (50) and Santachiara et al. (56) using mouse-human hybrids have shown a syntenic relationship between LDH-B and peptidase-B (PEP-B). The PEP-B/LDH-B synteny has been confirmed by Shows (58, 59), Van Cong et al. (66), and van Someren et al. (68). Jones et al. (23) using Chinese hamster–human hybrids have reported a syntenic relationship between LDH-B and the complement to the Chinese hamster glycine auxotrophic mutant A. Serine hydroxymethylase has been implicated as the specific deficiency in glycine A auxotrophy. Van Heyningen et al. (67) have reported a syntenic association between human mitochondrial MOR and mitochondrial citrate synthetase and LDH-B using man-mouse hybrids.

Chromosome 13

No assignment.

Chromosome 14. NP

Ricciuti and Ruddle (45, 46), using a 14/X translocation in a human diploid fibroblastic cell strain (KOP) hybridized to a mouse cell line, have demonstrated a segregation of nucleoside phosphorylase (NP) with the X-linked markers HGPRT, GPD, and PGK. Somatic cell genetic (46, 49) and family studies (12) have provided evidence for the autosomal linkage of NP. The studies of Ricciuti and Ruddle thereby support the assignment of NP to chromosome 14. Hamerton et al. (19) have reported results based on Chinese hamster–mouse hybrids which are consistent with the above findings of Ricciuti and Ruddle.

Chromosome 15

No assignment.

Chromosome 16. APRT

Tischfield and Ruddle *(unpublished data)* using an adenine phosphoribosyltransferase- (APRT) deficient mouse cell line hybridized to normal human-diploid cells have obtained evidence for the assignment of APRT to chromosome 16. Robson et al. (47) have assigned α-haptoglobin to chromosome 16 on evidence from family studies. APRT activity variants have been reported in man. It would be reasonable to identify kindreds in which α-haptoglobin and APRT variants are jointly expressed in order to test for linkage between these two markers.

Chromosome 17. TK

Green (16) and Migeon and Miller (38) using mouse-human cell hybrids assigned thymidine kinase (TK) to an E group chromosome. This assignment was based on the earlier findings of Weiss and Green (69). Boone and Ruddle (2) verified this assignment. Miller et al. (39), Ruddle and Chen (52), and Boone et al. (3) making use of a spontaneously occurring 17 translocation to a mouse chromosome have provided evidence for the assignment of TK to the long arm of chromosome 17. Kit et al. (26) and MacDougall (32) have demonstrated that adenovirus 12 infection induces host TK activity, and concomitantly a secondary constriction in the proximal segment of the long arm of 17. These findings suggest that the TK gene may be located near the adeno-12 induced uncoiler region.

Chromosome 18. PEP-A

Creagan et al. (11) using mouse-human cell hybrids have provided evidence for the assignment of peptidase A (PEP-A) to chromosome 18.

Chromosome 19. GPI

McMorris et al. (34) using mouse-human cell hybrids have provided evidence for the assignment of glucosephosphate isomerase (GPI) to chromosome 19. Hamerton et al. (19) have confirmed this assignment using Chinese hamster–human cell hybrids. Linkage studies in the mouse have revealed a loose linkage between GPI and β hemoglobin. It will be of interest to test for a similar linkage relationship in human kindreds.

Chromosome 20. ADA

Boone et al. (3) using mouse-human hybrids reported a weak association among cytoplasmic isocitrate dehydrogenase (IDH), cytoplasmic malate oxidoreductase (MOR), and chromosome 20. Tischfield, Creagan, and Ruddle *(unpublished data)* using more extensive data from mouse-human hybrids have shown that IDH and MOR cannot be assigned to 20, and using mouse-human hybrids they have obtained evidence for the assignment of adenosine deaminase (ADA) to chromosome 20. ADA is asyntenic with both IDH and MOR.

Chromosome 21. IPO-A—AVP

Tan et al. (62) using somatic cell hybrids have provided evidence for the syntenic association among indophenoloxidase-A, dimeric (IPO-A), and the antiviral response or the antiviral protein (AVP). AVP is the presumed protein factor which mediates specific interferon inhibition of virus replication. Tan et al. (62) have also shown that genetic factors responsible for interferon production and AVP synthesis are asyntenic. These results parallel earlier studies by Cassingena et al. (4) on their asyntenic association based on monkey-rat somatic cell hybrids. Tan et al. (62) have assigned AVP/IPO-A to chromosome 21.

Chromosome 22

No assignment.

Chromosome X. GPD—PGK—HGPRT—α-GAL

Glucose-6-phosphate dehydrogenase (GPD), hypoxanthine-guanine phosphoribosyltransferase (HGPRT), and phosphoglycerate kinase (PGK) have all been assigned to the X chromosome by segregation analysis in families (33). Nabholz et al. (41) using mouse-human hybrids confirmed the X linkage of HGPRT. Meera Kahn et al. (35) have demonstrated the X linkage of PGK by cell hybrid analysis. Ruddle et al. (53) using mouse-human hybrids confirmed the X linkage of HGPRT, glucose-6-phosphate dehydrogenase (GPD), and phosphoglycerate kinase (PGK). Grzeschik et al. (18) have recently provided evidence based on Chinese hamster–human hybrids for the assignment of α-galactosidase (α-Gal) to the X chromosome. In an earlier report, Grzeschik et al. (17) analyzed cell hybrids between human KOP cells which possess a 14/X translocation (KOP) and mouse and Syrian hamster cells. They observed an infrequent (5%) segregation of HGPRT and G6PD from PGK. This led them to postulate the assignment of HGPRT and GPD to the short arm, although assignment to the long arm was not altogether discounted. Ricciuti and Ruddle (46) using the same KOP material hybridized to mouse cells have obtained data which indicate that all three markers are located on the X long arm. Ricciuti and Ruddle also propose that PGK is proximal to the centromere and distant from the other two markers. HGPRT and GPD are postulated to be close together and distal to the centromere with respect to PGK. Preliminary evidence would indicate that HGPRT is proximal to G6PD. Gerald and coworkers *(personal communication)* have recently studied a human cell with a 19/X translocation hybridized to mouse cells. A translocation product composed of the 19 short arm, the proximal half of 19 long arm, and the distal half of the X long arm was correlated with GPI, HGPRT, and GPD, but not PGK. This result is consistent with the human X-linkage map proposed by Ricciuti and Ruddle (46). It also confirms the assignment of GPI to 19.

Chromosome Y

No assignment.

Somatic cell genetics provides possibilities for new approaches to mammalian genetics. Holandric inheritance has not been established in man. The human Y chromosome is retained in 11% of 28 primary clones analyzed by our laboratory. Y-bearing hybrid clones provide very possibly a means of assigning genetic elements to the Y chromosome. Hybrid cells also provide an approach to the mapping of active and inactive regions of the X chromosome. It is conceivable that the human population carries an appreciable number of nulloalleles that do not code for an active gene product. Such alleles might not be detected if homozygosity for the nulloalleles resulted in early fetal death or if homozygosity produced only a slight reduction in the fitness of the homozygote. Nulloalleles could result from gene deletion, temperature instability of the gene product, or the interruption of transcription by the incorporation of termination codons. Somatic genetics may allow the detection of nulloalleles, first by establishing gene assignments to chromosomes, and second by detecting instances of chromosome presence associated with corresponding phenotype absence in hybrid cells formed from the cells obtained from particular carrier families.

The development of a detailed human genetic map is certain to provide insight into the evolutionary origins of man and the primates. LDH-A and LDH-B are located on chromosomes 11 and 12 respectively. These chromosomes are similar in size, centromere position, and banding pattern. This result is consistent with the occurrence of an ancestral polyploidization event in the early primate genome as discussed by Comings (9). Somatic cell genetic analysis should be feasible for representative members of the order Primates using rodent x primate hybrids. Linkage data from such hybrids should provide information on the relatedness of these forms and yield estimates for rates of evolutionary divergence, especially when combined with comparative studies on the chromosome constitutions and amino acid sequences of homologous proteins in the representative specimens.

It is already obvious that somatic cell genetics has and will in the future contribute importantly to human genetics. Moreover, these developments enhance the significance and future role of family and population genetic studies. Already map distances between genes have been refined or established by the certain knowledge from somatic cell genetics that particular gene pairs are syntenic. We should expect a fruitful interaction and collaboration between practitioners of somatic cell genetics and classical genetics. However, we must keep clearly in mind that somatic genetic procedures as they currently exist are still unacceptably slow and that precise linkage estimates can not yet be made. If we are soon to develop genetic maps of man comparable to those available for the lower eukaryotes, it will be necessary to develop new procedures. In this respect, interesting possibilities exist with regard to somatic recombination and gene transfer.

REFERENCES

1. Barski, G., Sorieul, S., and Cornefert, F. (1960): Production dans des cultures *in vitro* de deux souches cellulaires en association, de cellules de caractère "hybride." *C. R. Acad. Sci., Paris 251*:1825.
2. Boone, C. M., and Ruddle, F. H. (1969): Interspecific hybridization between human and mouse somatic cells: Enzyme and linkage studies. *Biochem. Genet. 3*:119.
3. Boone, C. M., Chen, T. R., and Ruddle, F. H. (1972): Assignment of three human genes to chromosomes (LDH-A to 11, TK to 17, and IDH to 20) and evidence for translocation between human and mouse chromosomes in somatic cell hybrids. *Proc. Nat. Acad. Sci. USA 69*:510.
4. Cassingena, R., Chany, C., Vignal, M., Suarez, H., Estrade, S., and Lazar, P. (1971): Use of monkey-mouse hybrid cells for the study of the cellular regulation of interferon production and action. *Proc. Nat. Acad. Sci. USA 68*:580.
5. Chasin, L. A. (1972): Non-linkage of induced mutations in Chinese hamster cells. *Nature New Biol. 240*:50.
6. Chen, T. R., McMorris, F. A., Creagan, R., Ricciuti, F., Tischfield, J., and Ruddle, F. H. (1973): Assignment of the genes for malate oxidoreductase decarboxylating to chromosome 6 and peptidase B and lactate dehydrogenase to chromosome 12 in man. *Am. J. Hum. Gen. 25*:200.
7. Chu, E. H. Y., Brimer, P., Jacobson, K. B., and Merriam, E. V. (1969): Mammalian cell genetics I. Selection and characterization of mutations auxotrophic for L-glutamine or resistant to 8-azaguanine in Chinese hamster cells *in vitro*. *Genetics 62*:359.
8. Chu, E. H. Y., Sun, N. C., and Chang, C. C. (1972): Induction of auxotrophic mutations by treatment of Chinese hamster cells with 5-bromodeoxyuridine and black light. *Proc. Nat. Acad. Sci. USA 69*:3459.
9. Comings, D. E. (1972): Evidence for ancient tetraploidy and conservation of linkage groups in mammalian chromosomes. *Nature 238*:455.
10. Creagan, R. P., Kucherlapati, R., Tischfield, J., Chen, S., Ricciuti, F., and Ruddle, F. H. (1973): Chromosome assignment in man of the genes for isocitrate dehydrogenase (IDH) and malate oxidoreductase (MOR). Abstract 24th Annual Meeting of the Tissue Culture Association, Boston.
11. Creagan, R., Tischfield, J., McMorris, F. A., Chen, S., Hirschi, M., Chen, T. R., Ricciuti, F., and Ruddle, F. H. *(in press):* Assignment of genes for human peptidase A to chromosome 18 and cytoplasmic glutamate oxaloacetate transaminase to chromosome 10 using somatic cell hybrids. *Cytogenetics*.
12. Edwards, Y. H., Hopkinson, D. A., and Harris, H. (1971): Inherited variants of human nucleoside phosphorylase. *Ann. Hum. Gen. 34*:395.
13. Ephrussi, B., and Weiss, M. C. (1965): Interspecific hybridization of somatic cells. *Proc. Nat. Acad. Sci. USA 53*:1040.
14. Ephrussi, B., and Weiss, M. C. (1967): Regulation of the cell cycle in mammalian cells: Inferences and speculations based on observations of interspecific somatic hybrids. In: *Control Mechanisms in Developmental Processes* (ed. M. Locke), Society for Developmental Biology Symposium No. 26, Devel. Biol. Suppl. I., p. 136.

15. German, J. (1964): Cytological evidence for crossing-over *in vitro* in human lymphoid cells. *Science 144*:298.
16. Green, H. (1969): Prospects for the chromosomal localization of human genes in human-mouse somatic cell hybrids. Heterospecific genome interaction. The Wistar Institute Symposium Monograph No. 9, pp. 51–59.
17. Grzeschik, K. H., Allderdice, P. W., Grzeschik, A., Opitz, J. M., Miller, O. J., and Siniscalco, M. (1972): Cytological mapping of human X-linked genes by use of somatic cell hybrids involving an X-autosome translocation. *Proc. Nat. Acad. Sci. USA 69*:69.
18. Grzeschik, K. H., Grzeschik, A. M., Banhof, S., Romeo, G., Siniscalco, M., Van Someren, H., Meera Khan, P., Westerveld, A., and Bootsma, D. (1972): X-Linkage of human α-galactosidase. *Nature New Biol. 240*:48.
19. Hamerton, J. L., Douglas, G. R., Gee, P., and Richardson, B. J. *(in press):* The association of glucose phosphate isomerase expression with human chromosome 19 using somatic cell hybrids. *Cytogenetics.*
20. Harris, M. (1964): *Cell Culture and Somatic Variation,* Holt, Rinehart, and Winston, New York.
21. Harris, M. (1973): Phenotypic expression of drug resistance in hybrid cells. *J. Nat. Cancer Inst. 50*:423.
22. Harris, H., and Watkins, J. F. (1965): Hybrid cells derived from mouse and man: Artificial heterokaryons of mammalian cells from different species. *Nature 205*:640.
23. Jones, C., Wuthier, P., Kao, F., and Puck, T. T. (1972): Genetics of somatic mammalian cells. XV. Evidence for linkage between human genes for lactic dehydrogenase B and serine hydroxymethylase. *J. Cell. Physiol. 80*:291.
24. Kao, F. T., and Puck, T. T. (1968): Genetics of somatic mammalian cells. VII. Induction and isolation of nutritional mutants in Chinese hamster cells. *Proc. Nat. Acad. Sci. USA 60*:1275.
25. Kao, F. T., and Puck, T. T. (1972): Genetics of somatic mammalian cells: Demonstration of a human esterase activator gene linked to the Ade B gene. *Proc. Nat. Acad. Sci. USA 69*:3273.
26. Kit, S., Nakajima, K., and Dubbs, D. R. (1970): Origin of thymidine kinase in adenovirus-infected human cell lines. *J. Virol. 5*:446.
27. Klein, D., Merchant, D. L., Klein, J., and Shreffler, D. C. (1970): Persistence of H-2 and some non-H-2 antigens on long-term-cultured mouse cell lines. *J. Nat. Cancer Inst. 44*:1149.
28. Lederberg, J. (1958): Genetic approaches to somatic cell variation: summary comment. *J. Cell. Physiol. 52*:383.
29. Littlefield, J. W. (1964): Selection of hybrids from matings of fibroblasts *in vitro* and their presumed recombinants. *Science 145*:709.
30. Littlefield, J. W. (1966): The use of drug-resistant markers to study the hybridization of mouse fibroblasts. *Exp. Cell Res. 41*:190.
31. McBride, O., Ozer, W., and Ozer, H. L. (1973): Transfer of genetic information by purified metaphase chromosomes. *Proc. Nat. Acad. Sci. USA 70*:1258.
32. McDougall, J. K. (1971): Adenovirus-induced chromosome aberrations in human cells. *J. Gen. Virol. 12*:43.
33. McKusick, V. A. (1971): *Mendelian Inheritance in Man,* The Johns Hopkins Press, Baltimore.
34. McMorris, F. A., Chen, T. R., Ricciuti, F., Tischfield, J., Creagan, R., and Ruddle, F. (1973): Chromosome assignments in man of the genes for two hexosephosphate isomerases. *Science 179*:1129.
35. Meera Khan, P., Westerveld, A., Grzeschik, K. H., Deys, B. F., Garson, O. M., and Siniscalco, M. (1971): X-Linkage of human phosphoglycerate kinase confirmed in man-mouse and man–Chinese hamster somatic cell hybrids. *Amer. J. Hum. Genet. 23*:614.
36. Meiss, H. K., and Basilico, C. (1972): Temperature sensitive mutants of BHK 21 cells. *Nature New Biol. 239*:66.
37. Merril, C. R., Geier, M. R., and Petricciani, J. C. (1971): Bacterial virus gene expression in human cells. *Nature 233*:398.
38. Migeon, B. R., and Miller, C. S. (1968): Human-mouse somatic cell hybrids with single human chromosome (group E): Link with thymidine kinase activity. *Science 162*:1005.
39. Miller, O. J., Miller, D. A., Kouri, R. E., Allderdice, P. W., Dev, V. G., Grewal, M. S.,

and Hutton, J. J. (1971): Identification of the mouse karyotype by quinacrine fluorescence, and tentative assignment of seven linkage groups (*Mus musculus*/autosomes/X and Y chromosomes). *Proc. Nat. Acad. Sci. USA 68*:1530.
40. Munyon, W., Kraiselburd, E., Davis, D., and Mann, J. (1971): Transfer of thymidine kinase to thymidine kinaseless L cells by infection with ultraviolet-irradiated herpes simplex virus. *J. Virology 7*:813.
41. Nabholz, M., Miggiano, V., and Bodmer, W. (1969): Genetic analysis with human-mouse somatic cell hybrids. *Nature 223*:358.
42. Naha, P. M. (1969): Temperature sensitive conditional mutants of monkey kidney cells. *Nature 223*:1380.
43. Pontecorvo, G. (1962): Methods of microbial genetics in an approach to human genetics. *Brit. M. Bull. 18*:81.
44. Puck, T. T., Wuthier, P., Jones, C., and Kao, F. (1971): Genetics of somatic mammalian cells: Lethal antigens as genetic markers for study of human linkage groups. *Proc. Nat. Acad. Sci. USA 68*:3102.
45. Ricciuti, F. and Ruddle, F. H. (1973): Assignment of nucleoside phosphorylase to D-14 and localization of X-linked loci in man by somatic cell genetics. *Nature 241*:180.
46. Ricciuti, F. C., and Ruddle, F. H. *(in press):* Assignment of three gene loci (PGK, HGPRT, G6PD) to the long arm of the human X-chromosome by somatic cell genetics. *Genetics.*
47. Robson, E. B., Polani, P. E., Dart, S. J., Jacobs, P. A., and Renwick, J. H. (1969): Probable assignment of the alpha locus of haptoglobin to chromosome 16 in man. *Nature 223*:1163.
48. Ruddle, F. H. (1972): Linkage analysis using somatic cell hybrids. In: *Advances in Human Genetics,* Vol. 3 (eds. H. Harris and K. Hirschhorn) Plenum Press, New York, pp. 173–235.
49. Ruddle, F. H. (1973): Linkage analysis in man by somatic cell genetics. *Nature 242*:165.
50. Ruddle, F. H., Chapman, V. M., Chen, T. R., and Klebe, R. J. (1970): Linkage between human lactate dehydrogenase A and B and peptidase B. *Nature 227*:251.
51. Ruddle, F. H., and Nichols, E. A. (1971): Starch gel electrophoretic phenotypes of mouse x human somatic cell hybrids and mouse isozyme polymorphisms. *In Vitro 7*:120.
52. Ruddle, F. H., and Chen, T. R. (1971): Utilization of centric heterochromatin for chromosome identification in somatic cell hybrids. In: *Perspectives in Cytogenetics* (ed. S. W. Wright and B. F. Crandall) Charles Thomas, Colorado Springs.
53. Ruddle, F. H., Chapman, V. M., Ricciuti, F., Murnane, M., Klebe, R., and Meera Khan, P. (1971): Linkage relationships of seventeen human gene loci as determined by man-mouse somatic cell hybrids. *Nature 232*:69.
54. Ruddle, F. H., Ricciuti, F., McMorris, F. A., Tischfield, J., Creagan, R., Darlington, G., and Chen, T. R. (1972): Somatic cell genetic assignment of peptidase-C and the Rh linkage group to chromosome A-1 in man. *Science 176*:1429.
55. Ruddle, F. H., and Chen, T. R. (1973): Gene assignment in man by somatic cell genetics. In: *Genetics and the Skin,* Vol. 12, Cutaneous Genetics (eds. W. Montagna, L. M. Pasztor, W. Quevedo, Jr., and P. Porter), Appleton-Century Cross, New York; *J. Invest. Dermatol. 60.*
56. Santachiara, A. S., Nabholz, M., Miggiano, V., Darlington, A. J., and Bodmer, W. (1970): Genetic analysis with man/mouse somatic cell hybrids: Linkage between human lactate dehydrogenase B and peptidase B genes. *Nature 227*:248.
57. Shaw, M. W., and Cohen, M. M. (1965): Chromosome exchanges in human leukocytes induced by mitomycin C. *Genetics 51*:181.
58. Shows, T. B. (1972a): Genetics of human-mouse somatic cell hybrids: Linkage of human genes for isocitrate dehydrogenase and malate dehydrogenase. *Biochem. Genet. 7*:193.
59. Shows, T. B. (1972b): Genetics of human-mouse somatic cell hybrids: Linkage of human genes for lactate dehydrogenase-A and esterase-A_4. *Proc. Nat. Acad. Sci. USA 69*:348.
60. Shows, T. B. (1972c): Linkage of loci for human pyruvate kinase and mannosephosphate isomerase in somatic cell hybrids. Abstr. *Am. J. Hum. Genet. 24*:31a.
61. Stern, C. (1958): The nucleus and somatic cell variation. *J. Cell. Comp. Physiol. 52*:1.
62. Tan, Y. H., Tischfield, J., and Ruddle, F. H. (1973): The linkage of genes for the human interferon induced anti-viral protein and indophenol oxidase-B traits to chromosome G-21. *J. Exp. Med. 137*:317.

63. Thompson, L. H., Mankovitz, R., Baker, R. M., Till, J. E., Siminovitch, L., and Whitmore, G. F. (1970): Isolation of temperature-sensitive mutants of L-cells. *Proc. Nat. Acad. Sci. USA 66*:377.
64. Thompson, L. H., Harkins, J. L., and Stanners, C. P. *(in press):* A mammalian cell mutant with a temperature sensitive leucyl-tRNA synthetase. *Proc. Nat. Acad. Sci. USA.*
65. Toniolo, D., Meiss, H. K., and Basilico, C. (1973): A temperature-sensitive mutation affecting 28 S ribosomal RNA production in mammalian cells. *Proc. Nat. Acad. Sci. USA 70*:1273.
66. Van Cong, N., Billardon, C., Picard, J. Y., Feingold, J., and Frézal, J. (1971): Laison probable (linkage) entre les locus PGM-1 et Pep-C chez l'homme. *C. R. Acad. Sci. 272*:485.
67. Van Heyningen, V., Craig, I., and Bodmer, W. (1973): Genetic control of mitochondrial enzymes in human-mouse somatic cell hybrids. *Nature 242*:509.
68. Van Someren, H., Meera Khan, P., Westerveld, A., and Bootsma, D. (1972): Two new linkage groups in man, both carrying different loci for lactate dehydrogenase and glutamic pyruvic transaminase. *Nature New Biol. 240*:221.
69. Weiss, M. C., and Green, H. (1967): Human-mouse hybrid cell lines containing partial complements of human chromosomes and functioning human genes. *Proc. Nat. Acad. Sci. USA 58*:1104.
70. Westerveld, A., and Meera Khan, P. (1972): Evidence for linkage between human loci for 6-phosphogluconate dehydrogenase and phosphoglucomutase$_1$ in man/Chinese hamster somatic cell hybrids. *Nature 236*:30.

DISCUSSION

Thompson: Would you comment on the problem of position effects in utilizing translocations to do gene mapping?

Ruddle: There might be some modulation of gene expression induced by the translocation, but this can be controlled if it proves to be a problem by examining the expression of genes in the parental cells prior to hybridization. It might be a problem especially in the translocation mapping that I discussed, but again I think one could determine that by merely examining the parental cells before hybridization.

Littlefield: You mentioned gene-dosage data for the antiviral response and chromosome 21, but how about dosage of indophenol oxidase A? Do you have any data on this yet?

Ruddle: Not yet, but we are working on that. Tischfield is working on this in Charles Epstein's laboratory. It is a hard problem to solve because we recognize that there are two forms of the indophenol oxidase present. One has to separate the two forms before one can carry out quantitative studies. This is being done. There appears to be a quantitative relationship as one might expect, but the work is not yet finished.

Pontecorvo: Did I understand you correctly that, on your back selection for the TK translocation, you got only partial loss, never the loss of the whole chromosome?

Ruddle: In some cases we get loss of the whole chromosome. At least we can't recognize any remnant of the original chromosome. In other cases we get just the partial loss of what we presume to be the long arm of 17.

Pontecorvo: Is that a 2S mouse hybrid? I saw a lot of chromosomes.

Ruddle: Yes. It is a 2S mouse hybrid.

Pontecorvo: How many chromosomes does it have all together?

Ruddle: It would be over 100. All of those were 2S and difficult to work with.

Pontecorvo: Were there three parents?

Ruddle: No. The mouse parent was RAG.

Migeon: Have you made any calculations or thought about the number of clones that one has to examine before one can conclude with confidence that two loci are indeed linked.

Ruddle: That is always a difficult question to resolve. In these instances, we have for all of these assignments looked at more than 20 clones. Generally, we look at results from a number of independent crosses, using different mouse-human inputs, and we find that there is agreement between the different crosses. We feel confident when we have very good correlation between the particular traits using 20 or 50 clones, or in some instances, maybe 100 clones. One can also make subclones and determine whether the relationship holds true in subclones. We set our standards for acceptance at a high level, demanding concordance at the 90% level. We also attempt to rationalize discordant clones on the basis of retesting chromosome constitutions and phenotypic expression. These checks are frequently applied to derivative subclones.

Somatic Cell Hybridization, edited by
Richard L. Davidson and Felix de la Cruz.
Raven Press, New York © 1974.

Somatic Cell Genetics of Enzyme Markers Associated with Three Human Linkage Groups*

Thomas B. Shows

Roswell Park Memorial Institute, New York State Department of Health, Buffalo, New York 14203

INTRODUCTION

Mapping the human genome now appears feasible when approached through a combination of classical family studies and parasexual somatic cell hybrid studies. The somatic cell hybrid approach has clearly led to the prediction and confirmation, in a very short time, of several human gene assignments (11). Employing the cell hybrid methodology and particularly the fact that human chromosomes and enzymes are lost in proliferating man-mouse cell hybrids, we have studied the linkage relationships of human enzyme markers associated with three linkage groups. The linkage groups, identified by their enzyme markers, are (a) pyruvate kinase-3 and mannose-phoshate isomerase, PK-3-MPI; (b) isocitrate dehydrogenase and malate dehydrogenase IDH-MDH; and (c) the X-linked markers hypoxanthine-guanine phosphoribosyl transferase, glucose-6-phosphate dehydrogenase, and phosphoglycerate kinase, HGPRT-G6PD-PGK.

The formation of cell hybrids, their isolation and proof of hybridization, and electrophoretic techniques have been described (15, 16). Other methodology will be presented in the table and figure legends.

THE MPI-PK-3 LINKAGE

A linkage relationship has been observed between PK-3 and MPI, both of which function in carbohydrate metabolism (17). Pyruvate kinase activity exists in human tissues as several different electrophoretic forms presumably under different genetic control. Gel electrophoresis distinguishes a red cell type (PK-1) which is associated with hemolytic anemia; a white cell type (PK-3) which is also present in other tissues and cultured fibroblasts; and perhaps a kidney type (PK-2) (1). MPI is present in tissues as a single molecular form after electrophoresis. The electrophoretic patterns of PK-3 and MPI in human diploid WI-38 fibroblasts, mouse L-cells and two hybrids

* Supported in part by NIH grants HD 05196 and GM 20454 and NSF grant GB-39273.

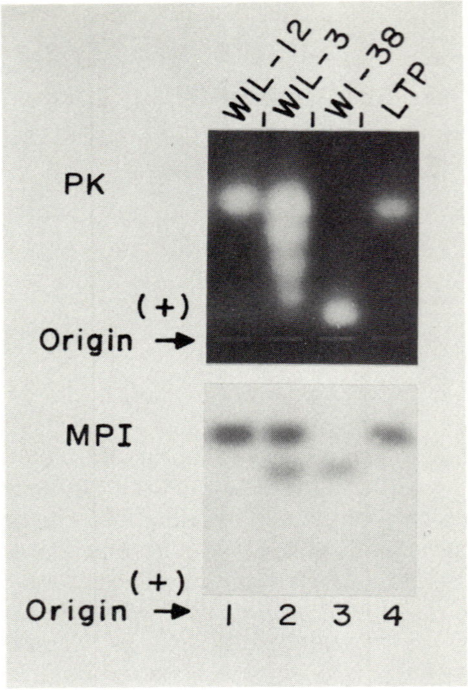

FIG. 1. Zymograms of PK-3 (top) and MPI (lower). Human fibroblast phenotypes (3); mouse L-cell phenotypes (4); human PK-3 and MPI positive hybrid, WIL-3 (2); and human PK-3 and MPI negative hybrid, WIL-12 (1). PK and MPI activity was separated by starch-gel electrophoresis as described by Shows and Ruddle (21) and PK and MPI activity visualized as reported by Bigley et al. (1) and Nichols et al. (9), respectively.

between these two parental lines are demonstrated in Fig. 1. In those hybrids which express human PK-3, MPI is also expressed as observed for hybrid WIL-3 (Fig. 1, channel 2). The PK phenotype, in cell hybrids which retain the human PK-3 gene, indicates that the enzyme is a tetrameric molecule since three hybrid enzymes are observed intermediate to the parental enzymes (Fig. 1, channel 2). No hybrid enzyme was observed for MPI.

The segregation characteristics of PK-3 and MPI in primary and secondary clones is observed in Table 1. There are four possible combinations of PK-3 and MPI expression in hybrids (+/+, +/–, –/+, –/–). If the two enzyme phenotypes are not coded by genes on the same chromosome, then all four combinations of enzyme expression are possible. But, if MPI and PK-3 are linked, then the markers are either expressed or absent together. Concordant segregation was observed for the two markers in primary and secondary clones with one exception, which is considered a result of chromosome breakage (Table 1). These results provide strong evidence for the

TABLE 1. *Human PK-3 and MPI segregation in man-mouse cell hybrids*

Primary clones[a]

		MPI	
		+	−
PK-3	+	13	1
	−	0	10

Secondary clones[b]

		MPI	
		+	−
PK-3	+	34	0
	−	0	17

[a] Primary clones were obtained from fusions of (a) mouse LTP and human WI-38 (15), (b) mouse A9 (HGPRT$^-$) X human TS-408, (c) mouse LM/TK$^-$ X human On Ser (HGPRT$^-$), and (d) mouse A9 (HGPRT$^-$) X human AnLy (19). The human cell lines WI-38 (CCL 75), TS-408, and On Ser (CRL 1112) are diploid fibroblasts. The AnLy line possesses an X/9 translocation. The mouse lines are L-cell derivatives and are selected against in HAT selection medium (5).

[b] Secondary clones were derived from 14 primary clones. The segregation data presented in the table was obtained by cloning primary clones that were positive for both markers. In addition, 36 secondary clones were derived from primary clones negative for both markers. All 36 clones were negative for MPI and PK-3 as expected.

linkage of genes that code for PK-3 and MPI. McMorris et al. (7) have found a positive correlation for MPI and chromosome 7. It is therefore reasoned that pyruvate kinase-3 is also located on chromosome 7.

It is important to note that this linkage relationship has only been observed in hybrids between L-cells and human fibroblasts. When RAG cells, derived from a mouse renal adenocarcinoma (4), were fused to human diploid fibroblasts, human mannosephosphate isomerase was expressed but human PK-3 was extinguished. Pyruvate kinase of L-cells differs in mobility from that of RAG pyruvate kinase. In RAG X L-cell hybrids the L-cell PK was extinguished. Similarly in L-cell X mouse neuroblastoma (Neuro-2ag) hybrids the L-cell PK was also extinguished. Therefore, in certain hybrid combinations, a regulatory mechanism is capable of suppressing the expression of a pyruvate kinase locus of one of the parental cells. The extinction of human PK-3 activity was attributed to suppression of activity rather than chromosome loss since MPI, on the same chromosome, was expressed.

THE IDH-MDH LINKAGE

The linkage of the cytoplasmic forms of human NADP-dependent isocitrate dehydrogenase (IDH) and NAD-dependent malate dehydrogenase (MDH, also referred to as MOR) (11) has been described previously (16). This observation was obtained in man-mouse hybrids and has been confirmed in another laboratory (11). Although the two phenotypes appear linked in man-mouse hybrids, this linkage has not been confirmed in Chinese hamster–man hybrids nor has the assignment to a specific human chromosome been established. Additional data will be presented here to support the linkage of the two citric acid cycle enzymes.

The segregation of IDH and MDH in hybrid clones from six different sets of cell hybrids using four different diploid human fibroblasts is demonstrated in Table 2. Of 62 primary clones, 47 expressed both human IDH and MDH, whereas in 11 primary clones both markers were absent. Four clones (6%) were discordant. In subclones isolated from IDH-MDH positive pri-

TABLE 2. *Human IDH and MDH segregation in man-mouse cell hybrids*

	IDH-MDH +/+	IDH-MDH +/−	IDH-MDH −/+	IDH-MDH −/−
Primary clones				
WIL (WI-38 X LTP)	6	0	1	6
REW (RAG X WI-38)	27	0	2	1
TSR (TS-408 X RAG)	6	0	1	1
TSA (TS-408 X A9)	0	0	0	2
LOS (LM/TK⁻ X On Ser)	1	0	0	1
ALA (AnLy X A9)	7	0	0	0
	47	0	4	11
Secondary clones				
WIL	42	0	0	34
TSR	21	0	1	3
LOS	2	0	0	6
	65	0	1	43

Hybridization and electrophoretic techniques have been reported (16). Cell lines were described in Table 1. RAG is a mouse HGPRT deficient line (4). Secondary hybrid clones were isolated from MDH-IDH positive primary clones. Three technical considerations should be noted when testing the enzymes in cell homogenates: (1) the human IDH component is sensitive to repeated freezing and thawing of cell extracts; (2) mouse and human MDH patterns possess several subbands that increase in staining intensity with storage of extracts and on electrophoresis in different buffer systems; in fact, under suboptimal conditions mouse subbands have been observed to migrate to the same locations as the human and hybrid enzyme components; (3) IDH activity can best be visualized in a pH 7.2 Tris-citrate buffer for starch gel electrophoresis and a 0.05 M Tris buffer, pH 8.5, for histochemical staining; these are modifications of the initial procedure (16).

mary clones, the two markers segregated together in 65 instances but were not expressed in 43 clones. There was only one discordant subclone in 109 subclones. All discordant clones expressed MDH but not IDH and could be explained if IDH is located at the end of a chromosome and is lost due to an occasional chromosomal breakage. Clones positive for IDH but negative for MDH have not been observed. We have observed a 4 to 6% rate of discordancy in primary clones for several different confirmed linkages, which is in agreement with rates observed by others. The data clearly indicate the linkage of cytoplasmic NADP-dependent IDH and NAD-dependent MDH in man-mouse cell hybrids.

THE HGPRT-G6PD-PGK X-LINKAGE

The segregation of human X-linked markers hypoxanthine-guanine phosphoribosyl transferase (HGPRT), glucose-6-phosphate dehydrogenase (G6PD), and phosphoglycerate kinase (PGK) has been examined in 233 primary and secondary hybrid clones isolated from fusions of four different diploid human fibroblast lines with mouse L-cell and RAG lines deficient in HGPRT. The hybrid clones were continuously grown on HAT (5) selection medium so that the retention of human HGPRT was always necessary for hybrid viability. Unless X-chromosome breakage occurs in these hybrids, all three markers are expected to be expressed (Fig. 2). If X-chromosome breakage and loss of the fragment occurs, then the sex-linked markers can be separated (8). When separation occurs, the combination of markers which remains in hybrids suggests the proximity of one marker to another and can thus predict the linear order of the genes involved. The success of determining the linear order in this way depends to a great extent on single breaks and loss of chromosomal fragments. It is likely, however, that double chromosome breaks and rearrangements occur since the cells are under continuous selection pressure to retain HGPRT for growth.

The electrophoretic phenotypes of human and mouse HGPRT, G6PD, and PGK are demonstrated in Fig. 2. Only human HGPRT was expressed in hybrids with no indication of a hybrid enzyme. The G6PD phenotype consisted of a mouse, human, and hybrid enzyme; the latter form confirmed hybridization. The PGK phenotype consisted of only mouse and human forms. The majority of hybrids retained all three human markers but two additional phenotypes were observed which did not include all the markers. When HGPRT is retained for viability in HAT medium, there are four phenotypes involving the three markers that can occur if the X-chromosome undergoes breakage (Table 3). Most hybrids (223) possessed all markers (Fig. 2, channels 5, 6) confirming their linkage. Three hybrids lost PGK but retained HGPRT and G6PD. This phenotype could occur by a break and loss of PGK or the translocation of a HGPRT-G6PD segment to another chromosome with subsequent loss of PGK (Fig. 2, channels 7, 8). Seven

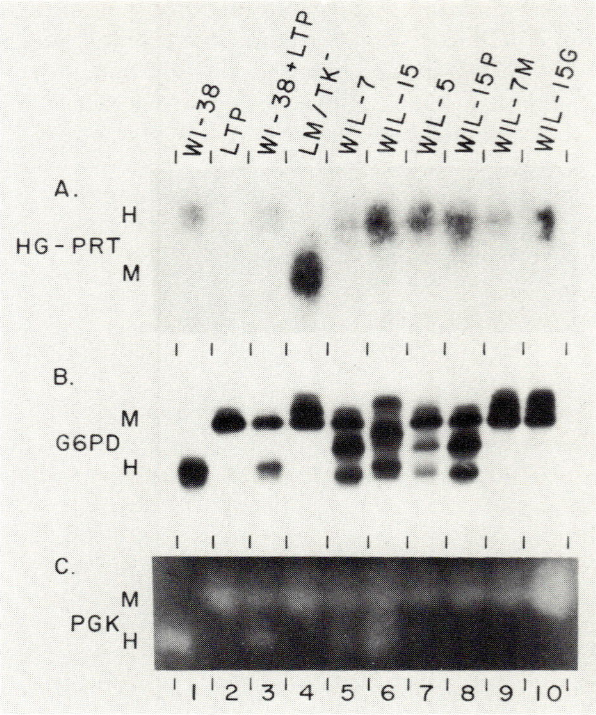

FIG. 2. Zymograms of HGPRT (A); G6PD (B); and PGK (C) phenotypes in human (channel 1), mouse (channels 2 and 4), and hybrid cells (channels 5 to 10). An artificial mixture of homogenates is demonstrated in channel 3; hybrids possessing all three markers in channels 5 and 6; hybrids possessing HGPRT and G6PD in channels 7 and 8; and hybrids possessing only HGPRT in channels 9 and 10. HGPRT was determined by the method of Shin et al. (14); G6PD electrophoresis and histochemical staining were accomplished by the methods of Shows and Ruddle (21) and Shows et al. (22), respectively; and PGK was analyzed according to Chen et al. (2).

clones lost G6PD and PGK (Fig. 2, channels 9, 10), but the possibility exists that a double break or complicated rearrangement occurred. The combination of HGPRT and PGK positive and G6PD negative did not occur among the small number of variant phenotypes observed (Table 3). The simplest observation regarding linear order (without chromosome studies) is that PGK and HGPRT are the outside markers in this triplet. The data that are being collected from the segregation of X-autosome translocations in cell hybrids support the observation that PGK is an outside marker but the translocation results suggest that G6PD is the other outside marker (3, 10, 12). We have not observed separation of X-linked markers in man-mouse hybrids that do not require human HGPRT for growth.

TABLE 3. *Separation of X-linked genes in cell hybrids*

HGPRT	G6PD	PGK	Hybrids
+	+	+	223
+	+	−	3
+	−	+	0
+	−	−	7

The hybrid clones were grown continuously on HAT selection medium. (+) Represents enzyme activity; (−) Indicates no activity. Seventy-eight primary clones were tested, the remainder were secondary clones. Five of the variant phenotypes were primary clones, five were secondary clones.

DISCUSSION

The assignment of enzyme markers to three linkage groups was described. Only the X-linkage of HGPRT, G6PD, and PGK has been confirmed in family studies. It is not likely that family studies will soon confirm the IDH-MDH and MPI-PK-3 linkages since genetic variants have not been reported to be segregating in man. While several linkages have been confirmed by family and cell-hybrid studies, as for example the X-linked markers (cf. refs. 6, 11, 19) and the markers (PGM_1, 6PGD, and PEP-C) assigned to chromosome 1 (cf. ref. 13), genetic variants do not exist for the majority of biochemical markers being tested in cell hybrids so that confirmation by Mendelian genetics is not imminent. A statistical analysis is therefore necessary to test the reliability of the linkage relationships determined in cell hybrids. A measure of the reliability of cell-hybrid linkage relationships can be estimated by fitting the enzyme segregation data to the Poisson distribution (16). This determines the probability of no markers per chromosome set, one marker per chromosome set, two markers per chromosome set, etc. The hybrid clones were anlyzed for 25 human enzyme markers that were segregating and the question was asked: what is the expected number of chromosome sets with no markers, 1 marker, 2 markers, etc. when there are 25 markers and 23 chromosome sets? For this calculation it was necessary to assume that human chromosomes (a) possessed the same number of genes, (b) segregated as whole structures, (c) were lost at random in hybrids, and (d) that human enzyme expression was not altered in hybrids.

The data are presented in Table 4. There were eight chromosomes expected with one marker (10 observed); five expected with two markers (six observed); and two linkage groups were expected to possess three markers (one was observed). Thus 17 linkage groups were observed (15 expected) with biochemical markers; six remained without markers (eight expected). Even with the assumptions listed above, the data conform closely to the expected distribution.

TABLE 4. *Poisson distribution of enzyme markers per chromosome set*

n Markers/chromosome set	*Pn*	Expected	Observed
0	0.336	8	6
1	0.366	8	10
2	0.199	5	6
3	0.072	2	1
4	0.019	0.4	0

$Pn = e^{-m} m^n/n!$. This equals the probability of *n* markers/chromosome set where there are 25 enzyme markers and 23 chromosome sets. The observed three-marker linkage was HGPRT-G6PD-PGK, and the 2 marker linkages were IDH-MDH, MPI-PK-3, PGM_1-PEP-C, LDH-A–EsA_4, LDH-B–PEP-B, and GOT-HK (18). The following markers are unlinked: ADA, AK-1, APRT, IPO-D, ME(MOD), NP, PEP-A, PEP-S, PHI, and hAP-1 (20).

This distribution of markers confirms the mounting evidence that human-gene linkage relationships, as determined by somatic cell hybrid methodology, represent a reliable alternative for mapping the human genome.

REFERENCES

1. Bigley, R. H., Stenzel, P., Jones, R. T., Campos, J. O., and Koler, R. D. (1968): Tissue distribution of human pyruvate kinase isozymes. *Enzym. Biol. Clin. 9*:10.
2. Chen, S. H., Malcolm, L. A., Yoshida, A., and Giblett, E. R. (1971): Phosphoglycerate kinase: An X-linked polymorphism in man. *Amer. J. Hum. Genet. 23*:87.
3. Grzeschik, K. H., Allderdice, P. W., Grzeschik, A., Opitz, J. M., Miller, O. J., and Siniscalco, M. (1972): Cytological mapping of human X-linked genes by use of somatic cell hybrids involving an X-autosome translocation. *PNAS 69*:69.
4. Klebe, R. J., Chen, T. R., and Ruddle, F. H. (1970): Mapping of a human genetic regulator element by somatic cell genetic analysis. *PNAS 66*:1220.
5. Littlefield, J. W. (1964): Selection of hybrids from matings of fibroblasts *in vitro* and their presumed recombinants. *Science 145*:709.
6. McKusick, V. A. (1971): *Mendelian Inheritance in Man,* John Hopkins Press, Baltimore.
7. McMorris, F. A., Chen, T. R., Ricciuti, F., Tischfield, J., Creagan, R., and Ruddle, F. (1973): Chromosome assignments in man of the genes for two hexosephosphate isomerases. *Science 179*:1129.
8. Miller, O. J., Cook, P. R., Kahn, P. M., Shin, S., and Siniscalco, M. (1971): Mitotic separation of two human X-linked genes in man-mouse somatic cell hybrids. *PNAS 68*:116.
9. Nichols, E. A., Chapman, V. M., and Ruddle, F. H. (1973): Polymorphism and linkage for mannose phosphate isomerase in *Mus musculus. Biochem. Genet. 8*:47.
10. Ricciuti, F., and Ruddle, F. H. (1973): Assignment of nucleoside phosphorylase to D-14 and localization of X-linked loci in man by somatic cell genetics. *Nature New Biol. 241*:180.
11. Ruddle, F. H. (1973): Linkage analysis in man by somatic cell genetics. *Nature 242*:165.
12. Ruddle, F. H. (1974): This volume.
13. Ruddle, F. H., Ricciuti, F., McMorris, F. A., Tischfield, J., Creagan, R., Darlington, G., and Chen, T. (1972): Somatic cell genetic assignment of peptidase C and the Rh linkage group to chromosome A-1 in man. *Science 176*:1429.
14. Shin, S., Kahn, M., and Cook, P. R. (1971): Characterization of hypoxanthine-guanine phosphoribosyl transferase in man-mouse somatic cell hybrids by an improved electrophoretic method. *Biochem. Genet. 5*:91.
15. Shows, T. B. (1972a): Genetics of human-mouse somatic cell hybrids: linkage of human genes for lactate dehydrogenase-A and esterase-A_4. *PNAS 69*:348.

16. Shows, T. B. (1972b): Genetics of human-mouse somatic cell hybrids: linkage of human genes for isocitrate dehydrogenase and malate dehydrogenase. *Biochem. Genet.* *7*:193.
17. Shows, T. B. (1972c): Linkage of loci for human pyruvate kinase and mannosephosphate isomerase in somatic cell hybrids. *Amer. J. Hum. Genet.* *24*:13a.
18. Shows, T. B. (1973): Linkage of human genes for glutamic oxalacetic transaminase and hexokinase in somatic cell hybrids. *New Haven Conference 1973: First International Workshop on Human Gene Mapping,* Birth Defects: Original Article Series IX: No. 6, 129. The National Foundation–March of Dimes, New York.
19. Shows, T. B., and Brown, J. A. (1973): X/9 Translocation suggests the assignment of G6PD, HGPRT and PGK to the long arm of the X chromosome in somatic cell hybrids. *New Haven Conference 1973: First International Workshop on Human Gene Mapping,* Birth Defects: Original Article Series IX: No. 6, 132. The National Foundation–March of Dimes, New York.
20. Shows, T. B., and Lalley, P. A. (1974): Control of lysosomal acid phosphatase expression in man-mouse cell hybrids. *Biochem. Genet.* *11*:123.
21. Shows, T. B., and Ruddle, F. H. (1968): Function of the lactate dehydrogenase *B* gene in mouse erythrocytes: Evidence for control by a regulatory gene. *Proc. Nat. Acad. Sci.* *61*:574.
22. Shows, T. B., Tashian, R. E., Brewer, G. J., and Dern, R. (1964): Erythrocyte glucose-6-phosphate dehydrogenase in Caucasians: New inherited variant. *Science* *145*:1056.

DISCUSSION

Ruddle: Referring to Table 3, were the clones that retained HGPRT under selective conditions?

Shows: They were all grown under selective conditions, which is very important from the point of view of breakage.

Ruddle: That may bias the results because it may actually enrich for complicated breaks, for example, double breaks and rearrangements.

Shows: That's right, in which case it would completely obscure linear order, but then such breaks could contribute to localizing the enzyme markers to small chromosomal segments.

Pontecorvo: Has anybody here observed the retention of the human X chromosome where it is not necessary?

Shows: We have observed a few mouse LM (TK^-) x human hybrid clones that were grown in HAT that have retained the X-linked markers HGPRT, G6PD, and PGK.

Siniscalco: Not all the cells in general?

Shows: We don't know that; we haven't tested individual cells in the hybrid populations.

Siniscalco: Were they investigated immunologically as well? When one looks at the biochemical markers, it is on the lysate of the entire cell population.

Shows: We tested for the enzyme markers in homogenates of primary and secondary clonal populations that were originally derived from single cells. We have not looked extensively for the X chromosome cytologically.

Ruddle: We have also observed retention of the X chromosome without selection for HGPRT, in fact in several clones. In the KOP series, we have seen one clone that has an intact X chromosome, which must be the inactive

X that has been retained. Also, in a number of clones we see the retention of the Y chromosome.

Evans: We also have one case, that is one line, involving a fusion between a human lymphocyte with mouse A9 cells in which murine HGPRT had been turned on but the human X chromosome was still present and producing detectable human HGPRT (as well as G6PD and PGK) up to around the tenth passage. In later subcultures the presence of human enzymes could not be detected [Watson, B., Gormley, I., Gardiner, S., Evans, H. J., and Harris, H. (1972): *Exp. Cell Res.* *75*:401].

Migeon: Do the hybrids express both mouse and human HGPRT? If so, is there a heteropolymer?

Shows: I have never seen a convincing heteropolymer migrating between mouse HGPRT and human HGPRT in the starch-gel system that we use.

Migeon: You have had the situation where both are expressed?

Shows: Yes.

Migeon: Only the parental isozymes are present?

Shows: Yes, but several possibilities exist that could either prevent the formation or the visualization of hybrid enzymes.

Migeon: The reason I ask is that it is not yet known whether HGPRT has subunits. The presence of heteropolymers in hybrids would help answer this question. If the enzyme is a monomer, then intragenic complementation between $HGPRT^-$ mutants may not be possible.

Shows: That's right, but intragenic complementation should be possible if HGPRT is a dimeric enzyme as has been suggested [Arnold, W., and Kelley, W. (1971): *J. Biol. Chem.* *246*:7398]. HGPRT activity has been observed after microscale isoelectric focusing of homogenates of a man-mouse hybrid whose parental cells were both HGPRT deficient. The hybrid nature of this enzyme has not been proven but it does migrate between the normal parental enzymes [Roy, K., and Ruddle, F. (1973): *Biochem. Genet.* *9*:175].

Siniscalco: If I am not wrong, Watson et al. (cited above) have reported the occurrence of a hybrid HGPRT between mouse and human cells recently.

Shows: There was no evidence in that paper that a hybrid HGPRT enzyme was present when both human and mouse HGPRT were present. In their electrophoretic system there were several minor components migrating adjacent to the parental HGPRT bands. In hybrids possessing both parental types the minor components could confuse the visualization of a hybrid enzyme. The hybrid cell phenotype appeared to be a summation of the two parental phenotypes.

Gerald: We have some data that would argue for the orientation of HGPRT perhaps being distal to PGK and G6PD. As Ruddle has pointed out, a complicated rearrangement could occur. It is of interest that in our hands the break in the triplet also seems to occur so that HGPRT is separated from the other two.

Shows: Was this done with translocations?

Gerald: This occurs both with and without translocations.

Shows: I think there is still much to be learned from breakage and translocation data in cell hybrids concerning the linear order and expression of X-linked markers.

Somatic Cell Hybridization, edited by
Richard L. Davidson and Felix de la Cruz.
Raven Press, New York © 1974.

Hybridization and Somatic Cell Genetics

Theodore Puck

Eleanor Roosevelt Institute for Cancer Research, University of Colorado Medical Center, Denver, Colorado 80220

Somatic cell genetics was designed in order to bypass the need for mating procedures of the whole organism in the study of mammalian genetics.

With the availability of single cell plating, means for the production of chromosomal and single gene mutants, and cell hybridization, it became possible to carry out virtually all of the genetic operations that are possible in simpler organisms like *E. coli,* with the exception of gene recombination. There is little doubt that this, too, will be possible before long. It should be noted that cell hybridization makes possible overcoming some of the classical genetic barriers. Thus, true crosses can be accomplished in somatic cell genetics between different species, and even classes, so that much more enriched experimentation is often possible (3). I know of no other field of science that promises as much excitement and accomplishment in the next decades.

Our laboratory has concentrated on the use of the Chinese hamster cell in hybridization studies because of three important advantages. First is the speed of cell growth. The generation time of the standard culture is 10 to 12 hr. Second is the speed of loss of human chromosomes when Chinese hamster cells are hybridized with human cells under our specific experimental conditions. Stable hybrids containing virtually the entire genome of the Chinese hamster cell and as few as one or two human chromosomes can be isolated within 1 to 2 weeks. This facilitates experimentation and minimizes the danger of chromosomal translocation, which can confuse the experimental findings. Finally, there are only 20 chromosomes in the particular Chinese hamster ovary cell that has become our standard culture, so that cytogenetic analysis is simplified.

By means of the technique using BUdR and visible light, it is possible to obtain reasonably large numbers of auxotrophic mutants by simple procedures. As of now, we have 16 different auxotrophic mutants of the Chinese hamster ovary (CHO) cell that have been produced by treatment with standard mutagens followed by the BUdR light-isolation procedure. These mutants provide good genetic markers and efficient selection systems which make possible many clean genetic experiments. For example, the original wild-type CHO cell grows with virtually 100% plating efficiency regardless of the presence or absence of glycine. The glycine-deficient mutant, however, exhibits no growth whatever in the absence of glycine. When glycine is added to the medium, the mutant exhibits a plating efficiency approaching 100%.

The currently available auxotrophic mutants exhibit requirements for nutrilites like glycine, proline, thymidine, adenine, serine, and inositol. Single and multiple mutational markers can be introduced into the same cell, so that more complex selection operations can be undertaken (1, 4, 5, 7).

Hybridization makes possible determination of dominance or recessiveness of different alleles of a given gene (11). This classical genetic operation first carried out in whole organisms by Mendel can now be simply performed with somatic cells *in vitro*. Thus if one hybridizes a glycine-requiring CHO mutant with an adenine auxotroph in medium lacking both metabolites, hybrid colonies are obtained that are prototrophic for both substances. All of our auxotrophic mutants have been shown by this methodology to be recessive to their respective wild-type genes.

Another important application of the hybridization phenomenon is complementation analysis (6, 8, 10, 11). Given a series of mutants with the same phenotype, this procedure permits determination of which of these have the same and which have different genotypes. For example, in one series of experiments 13 different glycine-requiring auxotrophs of the CHO cell were isolated after treatment of the wild-type with a variety of different mutagenic agents. These mutants were then selected, two at a time, hybridized with each other, and the resulting hybrids tested for their ability to grow in the absence of glycine. It was found that the 13 different glycine-requiring mutants formed four different complementation groups. Such groups are identified by the fact that whenever cells from different complementation groups are fused, the resulting hybrids can grow in the absence of glycine whereas hybridization of cells from the same complementation groups results in hybrids that still require glycine for growth. Thus this series of glycine auxotrophs represents four different gene mutations, and it can be concluded that at least four different genes can mutate independently so as to produce a glycine-growth requirement. The first of these mutants, which we have designated glycine A, is deficient in the enzyme serine hydroxymethylase, which converts serine to glycine. The other mutants, which by now can be differentiated biochemically as well as by complementation analysis, appear to be deficient in different steps of tetrahydrofolate metabolism which is also required in the reaction synthesizing glycine. The biochemistry of this other mutant is under continuing study. Similarly we have found six different complementation groups among the mutants that require adenine.

Use of the Chinese hamster cell for hybridization is also advantageous because of the large number of its auxotrophic mutants that are already available and the possibility of increasing this number by further applica-

Puck: I know of no continuously cultivatable hybrids containing chromosomal material from both species obtained between cells of different phyla. However, Kao in our laboratory does have continuing cultures produced by fusion of the Chinese hamster cell and the chick that contain a specific chick chromosome plus the regular CHO-KI chromosomes and that display complementing biochemical behavior demonstrating clearly the presence and activity of contributions from both genomes [Kao, F. (1973): *Proc. Nat. Acad. Sci. 70*:2893].

Davidson: Dr. Pontecorvo, could you amplify your statement that mouse-chicken hybrids do not exist.

Pontecorvo: The heterokaryons, of course, do exist. They have been used very extensively by the Oxford group in very interesting cell physiology work. However, the only case of claimed hybrids that I know of is the paper of Schwartz et al. [Schwartz, A., Cook, P., and Harris, H. (1971): *Nature New Biol. 230*:5] that can be interpreted in all sorts of ways. There are no hybrids there, only a kind of cell that has now reacquired an enzyme that wasn't there before. How the enzyme has been reacquired is another question.

Coon: A graduate student in my laboratory, David Trisler, has been looking at a series of propagating hybrids between cells of a TK minus BHK line and chick erythrocytes and chick fibroblasts. While he has not been able to show that some of the microchromosomes that are present are of chick origin, he does have evidence that the thymidine kinase deficiency is repaired by the hybridization, and also, that glucose phosphate isomerase is present in the avian form as well as the hamster form. Although these cells do not have all the chromosomes of both parents, it appears that some of the genes of both parents are present. Three hybrids were shown to contain 5 to 8% chicken DNA by cRNA/DNA hybridization. The hybrids exhibited chicken specific antigens 3.5 months after fusion.

Pontecorvo: If it is confirmed, I will leave it at that.

Watkins: Dr. Pontecorvo, could you list some of the "all sorts of ways" of interpreting the Schwartz et al. paper.

Pontecorvo: They have never seen a true hybrid. No trace of chick chromosomes has ever been seen in those cells. The only argument for saying it is a hybrid is that the HGPRT activity is of the chick type (in electrophoretic mobility) instead of the mouse type. If there were another proof of hybrid formation, I would say it is all right.

Dr. Ephrussi: At a recent conference I had detailed talks about this subject with Peter Cook. In fact, he doesn't speak anymore about chick enzyme or toad enzyme (in crosses between toad and mammalian cells), instead he speaks all the time about "toad-like" or "chick-like" HGPRT. He was asked why, and the answer was that, by electrophoretic tests, there is not exact accord between the enzyme in the "hybrids" and the chick or toad enzyme. I looked very carefully at Barbara Migeon's results, and there seems to be complete coincidence between the human HGPRT and the enzyme in the hybrids, but this does not seem to be the case with the other work.

Migeon: Is there not a distinct difference between the electrophoretic migration of chick and mouse HGPRT? I was wondering, what is the problem?

Pontecorvo: The problem is I don't believe that there is a hybrid unless you have some independent argument to show that there is a hybrid.

Siniscalco: Actually, the electrophoretic system used by Cook and Harris does not show a clear difference between mouse and chick HGPRT. I know of some unpublished data of Shin and Klinger who recently have utilized immunological evidence in addition to electrophoretic evidence. When tested with the antimouse HGPRT, the enzyme which appears in the hybrid cells does not react, which provides supporting evidence that it is really chick enzyme.

Somatic Cell Hybridization, edited by
Richard L. Davidson and Felix de la Cruz.
Raven Press, New York © 1974.

Strategies for X-Chromosome Mapping with Somatic Cell Hybrids

Marcello Siniscalco*

Department of Genetics, Faculty of Science, University of Naples, Italy

INTRODUCTION

The hybridization of somatic cells has offered a powerful tool for an experimental approach to the mapping of the human genome (13, 18, 21, 24). Data on the discovery of new linkage groups and on their chromosomal assignment are pouring in with a speed never experienced before in human genetics. However, as it is to be expected when scientific information accumulates very fast, many of these data are undecisive or definitely controversial. A good example of this state of affairs is the confusion about the cytological mapping of the human X-chromosome of which—as has been discussed in this volume—there exist several different versions. My personal opinion is that we are at a stage when the main efforts should still be devoted to the scrutiny of the entire methodology of linkage analysis via somatic cell hybrids for its reliability for the mapping of the human genome. The purpose of the present report is that of emphasizing the need of using different experimental strategies depending upon the aim that one wishes to achieve: (a) the detection of new linkage groups and their chromosomal assignment, (b) the statistical or (c) the cytological mapping of a given human chromosome. I will limit myself to the controversy about the human X-chromosome.

MITOTIC SEPARATION OF LINKED GENES, DISRUPTIVE STRATEGY AND STATISTICAL MAPPING OF THE X-CHROMOSOME

I must emphasize that when this beautiful tool, the man-mouse somatic cell hybrid, became available for studying human linkage, my colleagues and I were interested more in splitting off a known linkage group in the hybrid cells rather than in establishing new ones (12, 21, 22). This attitude was justified by the obvious consideration that true linkage maps could only be established by this approach if "linked genes" could segregate from one another in the hybrid cells through mitotic crossing over or similar phenomena.

* Present address: Sloan-Kettering Institute for Cancer Research, New York, New York.

With this thought in mind we decided to screen for the presence of human HGPRT and G6PD (both X-linked) in a series of man-mouse cell hybrids that Dr. O. J. Miller had derived in Dr. Henry Harris's laboratory at Oxford by fusing HGPRT-deficient mouse cells (A9) with six different diploid human strains bearing a normal active X-chromosome and HGPRT activity. Since these hybrid cells had been isolated and continuously maintained for over 6 months in the HAT-selective medium, the expectation was that both the above mentioned human X-linked markers should be retained unless segregation occurred as a result of chromosome breakage in a position intermediate to the cytological location of the corresponding structural loci.

The first set of results from this investigation reported by Miller et al. (12) showed for the first time the "mitotic separation" of the above-mentioned human X-linked markers in two out of the six different man-mouse somatic cell hybrid lines (and even in a higher proportion of the clones derived from each of them) which were expected to retain the whole group of X-linked genes having been continuously exposed to HAT selection.

Lately the same set of man-mouse hybrid cell lines was screened for the presence of a third marker of the human X-chromosome: the structural gene for phosphoglyceratekinase (PGK) whose X-linkage had in the meantime been established beyond doubt by pedigree studies (4) as well as by somatic cell genetics (10). As can be seen in Table 1, three of the original stock hybrid lines had lost also this marker, two of them being the ones that had lost also human G6PD. On subcloning these lines (20 subclones for each of the six original stock hybrid lines) 52 of 120 subclones expressed all of the three X-linked markers, 42 expressed human HGPRT but lacked G6PD and PGK, and 23 lacked only human PGK. Only three subclones, all derived from the same hybrid stock line, were HGPRT-positive, and lacked human G6PD although still possessing human PGK (Table 1 and refs. 22 and 23). Chromosomal studies, performed on the six stock hybrid lines and on 30 HAT subclones derived from them (five from each), disclosed the presence of at least one normal human C-chromosome[1] only in those hybrid cell lines that possessed all three human X-linked markers. However, even among these, the unquestionable presence on an intact human C-chromosome could be established only in about 10 to 15% of the 50 metaphases examined from each hybrid stock line or subclone. Thus we could confidently conclude that the loss of human G6PD and/or PGK observed in the hybrid cells of the type described after long propagation in the HAT-selective medium must be due to the partial loss of the human X-chromosome, although retention of G6PD and PGK together with human HGPRT does not necessarily prove the persistence of an intact X-chromosome in the hybrid cell. The latter

[1] These studies were performed before the era of chromosome banding, thus the presence or absence of the human-X was established by screening metaphases for the presence or absence of a human chromosome of the C-group.

TABLE 1. *Mitotic separation of three X-linked markers in a series of man-mouse somatic cell hybrids*

	HAT—Mass cultures			HAT—Clones			
Hybrid Lines[a]	HGPRT	G6PD	PGK	HGPRT	G6PD	PGK	No. of Clones
A9-HRBC2	H	M Hy H	M H	H	M Hy H	M H	17
				H	M Hy H	M	3
A9-CF	H	M Hy H	M H	H	M Hy H	M H	20
A9/ADC	H	M Hy H	M H	H	M Hy H	M H	15
				H	M	M H	3
				H	M	M	2
A9/4XY	H	M Hy H	M	H	M Hy H	M	20
A9/HRBC1	H	M	M	H	M	M	20
A9/JW	H	M	M	H	M	M	20
						Totals	120

H = human; Hy = hybrid; M = murine.
[a] See ref. 12.

finding suggests, in addition, that besides chromosome loss, chromosome rearrangements may be another frequent outcome when somatic hybrid cells of the kind discussed are grown in HAT medium for a prolonged time.

We have given the name of "disruptive strategy" to the experimental approach that favors the breakage or rearrangements of human chromosomes, and in particular of the human X-chromosome in somatic cell hybrids. Although little is known about the precise mechanism by which these events take place and eventually lead to the mitotic separation of X-linked genes, we feel that the data reported in Table 2 suggest—at least on a tentative basis—the two following conclusions: (a) the three structural loci under discussion must necessarily be at an appreciable distance from one another along the human X-chromosome for one or more breaks to occur between them; (b) their most likely sequence along the human X-chromosome may be HGPRT-G6PD-PGK in view of the finding that the hybrid cells of the kind described can easily lose human PGK when still retaining human G6PD while the reverse is very rarely observed (Table 1 and Fig. 1).

Both these conclusions are now supported by the work of others. The first one is in agreement with the findings of Nyhan et al (14) and Chen et al (4), who respectively found a high incidence of meiotic recombinants in pedigrees that gave information on linkage comparisons between HGPRT and G6PD or between G6PD and PGK. Our second conclusion has just received confirmation from the very similar set of data that Shows (20) has presented on the mitotic separation of human X-linked genes in somatic

TABLE 2. *Strategies for chromosome mapping with somatic cell hybrids*

1. Inactivated Sendai virus-mediated fusion of $HGPRT^-$ mouse or hamster cells with *Type A human cells* (diploid, with normal X, $HGPRT^+$) or *Type B human cells* (diploid, with functionally active X-autosomal translocation, $HGPRT^+$).
2. One month after fusion event, hybrid pseudoclones are isolated in HAT medium.
3. Different strategies are now applied according to the aim of the experiment:

Conservative strategy for:	Disruptive strategy for:
X-Linkage detection, with human parental cells type A *Cytological mapping,* with human parental cells type B	*X-Chromosome statistical mapping* with human parental cells type A
(a) Promptly transfer ±20 pseudoclones from HAT to MEM. (b) Propagate in MEM for at least 1 month, to allow for random loss of HGPRT-bearing human chromosomes. (c) Split each pseudoclone into two mass cultures, one in HAT, the other in 8-aza-guanine. Cells with human HGPRT are separated from those without it (see footnote 2). (d) (i) With human parental cells type A: Screening for retention or loss of "test markers" versus that of known human X-linked markers will allow *detection of X-linkage.* (ii) With human parental cells type B: Screening for retention or loss of human HGPRT and other X-linked markers versus that of "translocation chromosomes" will allow cytological mapping of the X-chromosome and possibly of the autosome involved in the translocation. Accurate chromosome studies with banding techniques are essential in this case.	(a) Let hybrid pseudoclones grow for at least 3 months in HAT. (All those appearing in a common culture vessel form a "hybrid stock culture.") (b) Clone: from each of 10 hybrid stock cultures, isolate 20 clones in HAT medium and 20 clones in 8-aza-guanine medium. (c) Analyze mitotic segregation of the known X-linked markers into "half-linkage groups" for *statistical mapping.* Past experience has shown that chromosome studies are, in this case, difficult and scarcely informative, even when banding techniques are used.

cell hybrids of the same type as ours and kept under experimental conditions identical to those I described above.

CONSERVATIVE STRATEGY AND X-LINKAGE DETECTION

From what has been said so far, it is clear that on the one hand the breakage of a linkage group in man-mouse or man-hamster cell hybrids may be a precious accident for the statistical mapping of the human genome with particular reference to those chromosomes whose retention or loss can be

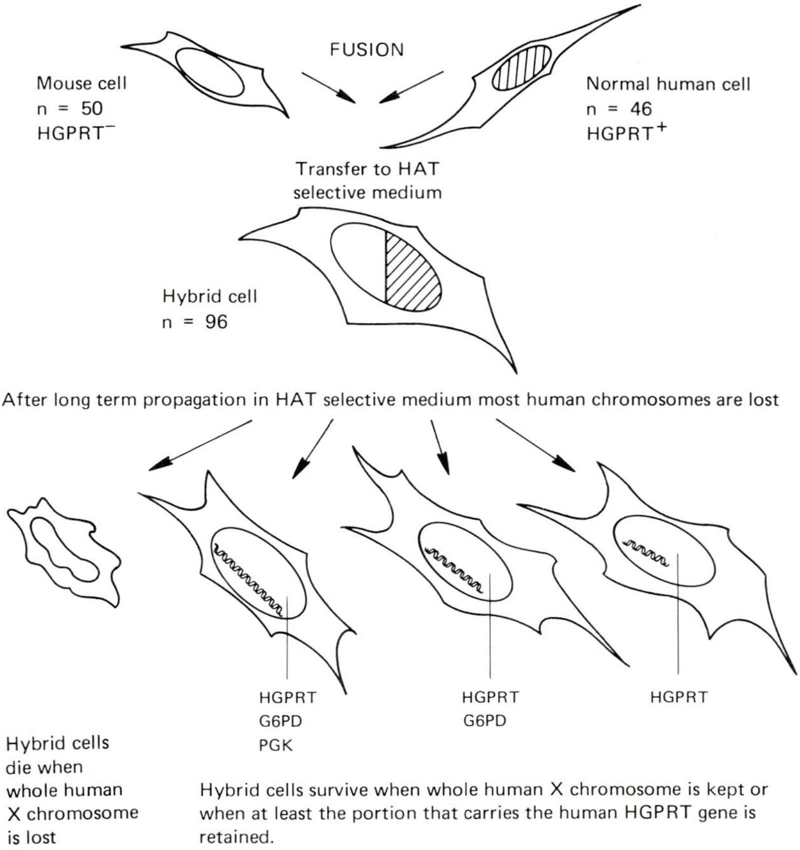

FIG. 1. Statistical mapping of the human genes for HGPRT, G6PD, and PGK along the X-chromosome, based on the mitotic separation of these X-linked markers reported in Table 1. Explanation in the text.

induced selectively in the hybrid genome [besides the X-chromosome, this can be done also for the human autosomes E16 (19) and E17 (11)]. On the other hand, it casts a legitimate doubt on the alleged superiority of the somatic cell hybrids of the type discussed over the classical methodology of pedigree analysis for the detection of "syntenic markers," i.e., those controlled by genes that, although located on the same chromosome, segregate independently at meiosis as a result of a high cross-over frequency (16).

Being just as much interested in the detection of new X-linked markers, we have investigated this problem very extensively. Thus we found that the mitotic separation of at least the X-linked markers can be practically avoided if hybrid cells are transferred to MEM (nonselective medium) soon after their isolation in HAT medium (i.e., within 2 to 3 weeks from the fusion experiment). The propagation in MEM will no longer force the retention of human HGPRT so that each hybrid pseudoclone will soon become a mosaic of hybrid cells *with* or *without* the human X that can be separated from one another by a short term massive subculturing[2] in the HAT and 8-azaguanine media respectively. Eventually the two kinds of hybrid cells are submitted to the biochemical and cytological studies required to establish whether or not a given "test marker" segregates together with the already known X-linked markers and with the X-chromosome.

The application of this procedure, which we call the "conservative strategy," has enabled us to confirm the X-linkage of the structural loci for human PGK (10) and alpha-galactosidase (7) since there was no exception to the "synteny" of these markers with human HGPRT and G6PD in several hundred hybrid clones that were derived from a series of different man-mouse and man-hamster hybrid cell lines produced through separate sets of fusion experiments.

An independent confirmation of our findings has been recently provided by Allderdice and Miller (2) who have used the technique of fluorescent banding to investigate the evolution of the karyotype in man-mouse somatic cell hybrids derived from Sendai virus mediated fusion of HGPRT–negative mouse cells with human cells bearing a normal X-chromosome. The part of their studies that is relevant to our own observations involves the cytological analysis of two hybrid sublines that were derived from the same hybrid pseudoclone soon after its isolation in HAT and were thereafter continuously propagated either in HAT or MEM medium.

These studies disclosed that, despite the fact that human HGPRT was essential for the survival of the subline grown in HAT, a normal human X-chromosome was observed in only 31% of the cells and the presence of a translocated segment of the X-chromosome could not be established. On the other hand, the subline that had been grown in the nonselective medium had a normal X-chromosome in as many as 20% of its cells.

In conclusion, we believe that the use of the "conservative strategy" is the condition required for the preservation of groups of "syntenic markers" in the interspecific hybrids of the type under consideration in man-mouse and man-hamster somatic cell hybrids. On the other hand, the "disruptive strategy" certainly favors the mitotic separation of human linkage and probably enhances the incidence of *de novo* chromosomal rearrangements in the whole genome of the hybrid cells.

---

[2] In performing this step one has to be careful to complete the selection of the HAT or 8-azaguanine resistant cells in the shortest time (to prevent *de novo* rearrangement) and to avoid the survival of HGPRT–negative hybrid cells by metabolic cooperation (15). This can be achieved by seeding at least 5×10^6 cells from each MEM pseudoclone into a large roller vessel.

STRUCTURAL CHROMOSOMAL ABERRATIONS AND CONSERVATIVE STRATEGY FOR THE CYTOLOGICAL MAPPING OF THE HUMAN GENOME WITH CELL HYBRIDS

In 1969 a well-documented case of X-autosomal reciprocal translocation $(t14q^{-}Xq^{+};\ tXq^{-}14q^{+})$ reported by Opitz and Pallister (15) made me wonder whether the hybridization of such cells with HGPRT–negative mouse or hamster cells would not represent a better system for the cytological mapping of the human genome and in particular of the human X-chromosome. I reasoned that the presence of a built-in reciprocal X-autosomal translocation in a somatic cell hybrid whose survival can be made dependent upon the retention or loss of human HGPRT, would turn out to be the choice tool at least for assigning to one or the other of the translocation products the X-linked markers displaced by the translocation event. For this to be possible, however, a number of conditions would have to be fulfilled:

(1) both or at least one of the translocation products would have to be functionally active in the parental as well as in the hybrid cells;
(2) both or at least one of the translocation products would have to be clearly identifiable in the hybrid genome;
(3) the "conservative strategy" would have to be rigorously applied to minimize the risk that any variation in the behavior of well-known human syntenic groups or isolated markers spotted in the hybrid cells might be due to *de novo* chromosomal rearrangements rather than to the inborn translocation carried by the human parental genome.

Dr. Opitz (15), using short-term leucocyte cultures, had already clearly established that the somatic cells of the heterozygous carrier of this reciprocal translocation—an otherwise normal woman—had the normal X-chromosome constantly inactivated. A mentally retarded male child born to this woman (actually the propositus who led to the discovery of the family) had instead an unbalanced karyotype including the reciprocal translocation of the maternal genome plus an additional dose of the $t14q^{-}Xq^{+}$ translocation product which turned out to be constantly inactivated. These findings were confirmed through later studies by Miller et al. (1) both on peripheral short-term lymphocyte cultures and on the fibroblastic cultures that we had set up, in the meantime, from primary skin explants obtained from the mother and the child for future use in cell hybrid studies. We called these two cell strains *KOP 1* and *KOP 2* respectively, and we decided to use the former—although they were both normal for all X-linked markers—in our own hybridization studies to avoid the confusion that might have arisen by using the latter, in view of the presence of one "active" and one "inactive" dose of the $t14\ q^{-}Xq^{+}$ translocation product.

The first set of data that my colleagues and I gathered with the use of this approach was very encouraging (8). We found that despite the rigorous application of the "conservative strategy," the set of somatic cell hybrids that we had derived from the Sendai virus mediated fusion of the *KOP 1 cells* with HGPRT–negative cells of mouse and Syrian hamster, yielding a large number of pseudoclones (over 10%) showing the loss of synteny between human PGK on the one hand and human HGPRT and G6PD on the other hand. In addition the loss or retention of human PGK appeared to be positively correlated with that of the $t14q^{-}Xq^{+}$ translocation chromosome. Thus, it was concluded that the structural locus for human PGK must be located on the distal part of the X-chromosome long arm, whereas the other two X-linked markers could be located anywhere along the remaining portion of the X-chromosome associated with the $tXq^{+}14q^{-}$ translocation product. These conclusions are included in Fig. 2. In this figure the proposed sequence of the three loci along the X-chromosome stems from the data referred to in the previous section and reported in full elsewhere (23).

We have seen in this volume that the use of human strains with reciprocal translocation is indeed becoming a very popular material for mapping the human genome via somatic cell hybridization. An array of different X-autosomal translocations has been utilized for this purpose by different groups of cell hybridizers. Unfortunately the conclusions drawn by each of these groups regarding the cytological map of the X-chromosome differ from ours and among themselves (3, 6, 17). This would be an interesting and valid finding pointing to the possible existence of X-chromosomal inversions in natural human populations or at least among the carriers of X-autosomal translocations—if other explanations can be ruled out. I fear there may be more trivial reasons for the controversial data, such as either failure to use

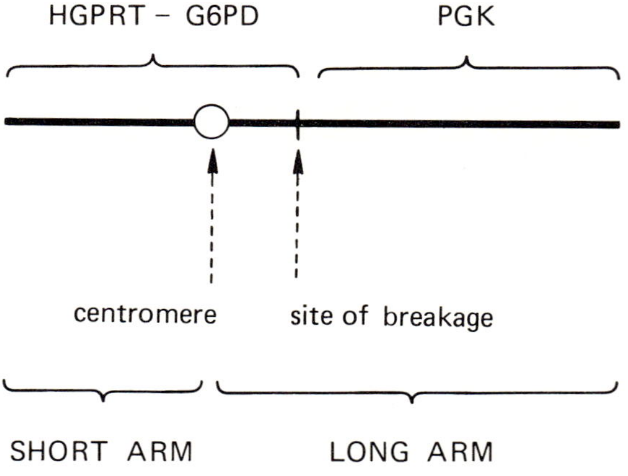

FIG. 2. Statistical and cytological map of the human X-chromosome based on the data of Table 1 and those reported in full in ref. 8.

the correct methodology or a spontaneous tendency to breakage on the part of translocation chromosomes, even when the conservative strategy is rigorously applied.

My suspicion arises particularly from the discordance between our own results and those of Ricciuti and Ruddle (17), existing despite the fact that the strains used as human parental cells (the KOP 1 strain by us and KOP 1 and 2 by them) carried the same X-autosomal translocation. A close look at Ricciuti and Ruddle's data, which they interpreted as suggesting the regular "synteny" of all human X-linked markers with the $t14q^{-}Xq^{+}$ translocation, reveals, however, that at least 5% of their hybrid clones must have undergone mitotic separation of the X-linked markers even if their interpretation is correct. Interpreting their data is further complicated by the fact that the KOP 2 cells were also utilized as a human parental strain (these cells carry, as I said earlier, two doses of the $t14q^{-}Xq^{+}$ translocation chromosome, thus making it difficult to establish the correlation between the retention of this chromosome and that of the X-linked markers).

An additional outcome of these studies has been the demonstration that human nucleoside phosphorylase (NP) segregates syntenically with the $t14q^{-}Xq^{+}$ translocation product, suggesting that the corresponding structural gene is located on chromosome D 14. Some unpublished observations of ours (9) are in full agreement with this conclusion.

To my mind, controversial results obtained with the use of the same original material convey an important message, i.e., the need to repeat (possibly in more than one laboratory) the whole set of critical experiments using all available strains with X-autosomal translocations, under conditions that minimize the occurrence of *de novo* chromosome rearrangements; this we have shown to be afforded by the use of our "conservative strategy," at least in the case of the interspecific somatic cell hybrids that have a human parental genome with a normal X chromosome. In addition, the "disruptive strategy" should be used intentionally for the purpose of comparison. Toward this end, I have immodestly spelled out, in Table 2, a summary of the different strategies to be used in linkage studies with somatic cell hybrids. Should they prove eventually to be worthless for the control of *de novo* chromosome rearrangements in cultured hybrid somatic cells with or without translocations, then we will have to fear that most of the conclusions on new "syntenic" groups and their chromosomal assignments arrived at thus far by means of somatic cell hybrids may be wrong, just as well.

ACKNOWLEDGMENT

I am indebted to Dr. Mathilde Krim for her advice and criticism on the final version of this manuscript.

REFERENCES

1. Allderdice, P. W., Miller, O. J., Klinger, H. P., Opitz, J. M., and Pallister, P. D. (1971): Demonstration of a spreading effect in an X-autosome translocation by combined autoradiographic and quinacrine-fluorescent studies. *Proc. 4th Int. Congr. Human Genetics* 14–15.
2. Allderdice, P. W., and Miller, O. J. (1973): Human chromosomes in 18 man-mouse somatic hybrid cell lines analyzed by quinacrine-fluorescence. *J. Cell Sci. 12*:809–830.
3. Bootsma, D. (1974). This volume.
4. Chen, F. H., Malcolm, L. A., Yoshida, A., and Giblett, E. R. (1971): Phosphoglycerate-kinase: An X-linked polymorphism in man. *Amer. J. Human Genetics 23*:87–91.
5. Fujimoto, W. Y., Subak-Sharpe, J. H., and Seegmiller, J. E. (1971): Hypoxanthine guanine phosphoribosyl transferase deficiency: Chemical agents selective for mutant or normal cultured fibroblasts in mixed and the heterozygote cultures. *Proc. Nat. Acad. Sci. USA 68*:1516–1519.
6. Gerald, P. S. (1974): This volume.
7. Grzeschik, K. H., Grzeschik, A. M., Benoff, S., Romeo, G., Siniscalco, M., van Someren, H., Meera Khan, P., Westerveld, A., and Bootsma, D. (1972): X-linkage of human α-galactosidase. *Nature New. Biol. 240*:48–50.
8. Grzeschik, K. H., Allderdice, P. W., Grzeschik, A., Opitz, J. M., Miller, O. J., and Siniscalco, M. (1972): Cytological mapping of human X-linked genes by use of somatic cell hybrids involving an X-autosome translocation. *Proc. Nat. Acad. Sci. USA 69*:69–73.
9. Grzeschik, K. H., and Siniscalco, M. *(in press):* Data quoted in the report of the subcommittee on the X-chromosome. *New Haven Conference 1973: First International Workshop on Human Gene Mapping,* Birth Defects: Original Article Series IX: No. 6. The National Foundation–March of Dimes, New York.
10. Meera Khan, P., Westerveld, A., Grzeschik, K. H., Deys, B. F., Garson, O. M., and Siniscalco, M. (1971): X-linkage of human phosphoglycerate kinase confirmed in man-mouse and man-Chinese hamster cell hybrids. *Amer. J. Human Genetics 23*:614–623.
11. Miller, O. J., Allderdice, P. W., Miller, D. A., Breg, W. R., and Migeon, B. R. (1971): Human thymidine kinase gene locus: assignment to chromosome 17 in a hybrid of man and mouse cells. *Science 173*:244–249.
12. Miller, O. J., Cook, P. R., Meera Khan, P., Shin, S., and Siniscalco, M. (1971): Mitotic separation of two human X-linked genes in man-mouse somatic cell hybrids. *Proc. Nat. Acad. Sci. USA, 68*:116–120.
13. Nabholz, M., Miggiano, V., and Bodmer, W. (1969): Genetic analysis with human-mouse somatic cell hybrids. *Nature 223*:358–363.
14. Nyhan, W. L., Bakay, B., Connor, J. D., Marks, J. F., and Keele, D. K. (1970): Hemizygous expression of glucose-6-phosphate-dehydrogenase in erythrocytes of heterozygotes for the Lesch-Nyhan syndrome. *Proc. Nat. Acad. Sci. USA 65*:214–218.
15. Opitz, J. M., and Pallister, P. D. (1969): Data presented at the annual meeting of the American Society of Human Genetics in 1969 but never published in full.
16. Renwick, J. H. (1971): The mapping of human chromosomes. *Ann. Rev. Genet. 5*:81–120.
17. Ricciuti, F., Ruddle, F. H. (1973): Assignment of nucleoside phosphorylase to D-14 and localization of X-linked loci in man by somatic cell genetics. *Nature New Biol. 241*:180–182.
18. Ruddle, F. H. (1970): Utilization of somatic cells for genetic analysis: possibilities and problems. *Symp. Int. Soc. Cell Biol. 9*:233–264.
19. Ruddle, F. H. (1972): Gene expression and gene regulation in mammalian cells. Paper presented at the ESGH Symposium, Amsterdam, September 23–24, 1972.
20. Shows, T. (1974): This volume.
21. Siniscalco, M. (1970): Somatic cell hybrids as tools for genetic studies in man. *Symp. Int. Soc. Cell Biol. 9*:205–231.
22. Siniscalco, M. (1972): Correction of genetic defects in cultured mammalian cells. *Adv. Biosci. 8*:307–327.
23. Siniscalco, M., Shin, S., Kooy-Veltknap, C., Meera Khan, P., Miller, O. J. *(submitted for publication):* Further data on X-linked genes in somatic cell hybrids.

24. Weiss, M. C., and Green, H. (1967): Human mouse hybrid cell lines containing partial complements of human chromosomes and functioning human genes. *Proc. Nat. Acad. Sci. USA 98*:1104–1111.

DISCUSSION

Davidson: In view of the recent results demonstrating the reappearance of HGPRT in so-called "nonreverting" lines, I would like to ask whether the claim that human HGPRT is present in hybrids is based on the demonstration of the human form of the enzyme or is this based only on growth in HAT.

Siniscalco: We have always looked for the enzyme electrophoretically.

Davidson: Has this been done by other people who are studying the X-chromosome? What effect could this have on linkage assignments?

Siniscalco: I suppose that everybody who works on linkage takes that into consideration.

Gerald: Are there any hybrids between human cells and HGPRT deficient rodent cells, which have been shown to be hybrids through studies of non-X-linked enzymes or chromosome markers, in which the HGPRT is of the rodent type? The clone described by Watson et al. which contained only rodent HGPRT was not a hybrid cell. I don't know of any case in which a hybrid cell defined as such by independent criteria had rodent HGPRT. Do you know of any?

Siniscalco: No.

Evans: The paper you refer to is the one describing joint work carried out in my own and Harry Harris' laboratories [Watson, B., Gormley, I., Gardiner, S., Evans, H. J., and Harris, H. (1972): *Exp. Cell Res. 75*:401] and the question you ask is whether we had any hybrids showing murine HGPRT as well as any non-X-linked human markers. We did not have any such cells, although the only non-X-linked locus that we were testing for was nucleoside phosphorylase and this was absent in the one cell line expressing the three X-linked markers, HGPRT, G6PD, and PGK of both human and murine origin.

Migeon: Watson et al. have some evidence that suggests that, in a fusion situation, A-9 will revert with a greater frequency than was previously thought. They also had one hybrid with murine in addition to human HGPRT and this hybrid had human chromosomal and G6PD markers as well.

Siniscalco: Actually, the data just mentioned are in agreement with what we described in a recent paper [Shin, S., Canera, R., Shildkraut, C., Klinger, H., and Siniscalco, M. (1973): *Nature New Biol. 241*:194]. We could find the revertants only after the fusion treatment. However, we never found the event at such a high frequency as reported in the paper from Harris' group.

Pontecorvo: I should like to repeat here what I said 2 years ago at a meeting of this nature, i.e., that you do get HGPRT reversion. In hamster-mouse hybridizations, there is a small proportion of cases with no trace of chromosomes of the other species. The revertants of the Chinese hamster are always of higher ploidy than the parental cells.

If you test the HGPRT minus cells by themselves you do not get the revertants, but I would not jump to the conclusion that there is something that the hybridization provides in terms of genetic information. My hypothesis is that the hybridization provides a feeder layer or something of this sort, which allows the revertants to become established while they do not become established when they are in pure culture.

Siniscalco: I entirely agree, especially as there is no need to use the virus. This seems to be a very good explanation.

Bootsma: We have studied hybrids between human white blood cells with an X-autosome translocation and TK^- Chinese hamster cells. There was no selective pressure on the X-chromosome in this case. It is interesting that the only useful fusions were obtained with the TK^- cells. We had previously hybridized these white blood cells with $HGPRT^-$ Chinese hamster cells and found splitting up of the X-chromosome. However, we could not find the piece of the X-chromosome in these hybrids. In the studies I will discuss the chromosome analysis was done by Pearson and the electrophoresis by Meera Khan and van Someren.

The breakpoint was in the long arm of the X-chromosome (see Discussion Fig. 1). A precise description of the translocation would be *t* (X*q*; 3*q*) (26; 13). The new element with the centromere of the number three chromosome is denoted 3/X while the deleted X is called the X/3 chromosome. We isolated three primary clones which had lost G6PD and retained PGK and HGPRT. These clones retained the X/3 marker indicating that G6PD is not on the short arm of the X. Two clones were isolated that had lost PGK and HGPRT and retained G6PD. Both had retained the 3/X marker chromosome. In the fusion of the white blood cells from the male carrier of the translocation, we found two clones that had retained PGK, HGPRT, and G6PD. Both clones had retained the X/3 and the 3/X chromosomes. One clone was isolated that had lost G6PD and retained PGK and HGPRT. However, we could find neither the X/3 nor the 3/X chromosome.

Although these results are preliminary, they indicate that G6PD is on the distal part of the long arm of the X, whereas the PGK and HGPRT loci would be on the short arm or the proximal part of the long arm of the X-chromosome.

Ruddle: Where is the region for PGK and HGPRT? It could be that our break point on the X is closer to the centromere than the break point in your translocation.

Bootsma: Yes.

Siniscalco: As I said, when different translocations are involved, it may not be surprising that different results are obtained. There is, however,

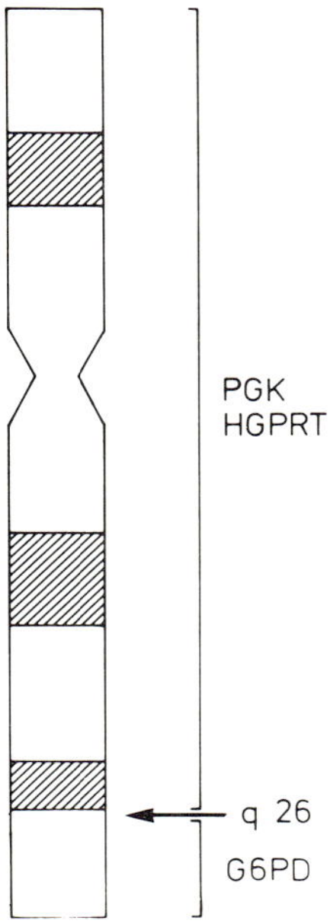

FIG 1. Schematic representation of the human X-chromosome showing the main bands and the break point which is found in the translocation involving the chromosomes X and 3.

another reason why we feel that the location of G6PD at the very tip of the chromosome is unlikely, based on a series of Klinefelter families from Sardinia where the mothers were heterozygous for G6PD. In a group of 15 such families in which the nondisjunction was established to be maternal on the basis of the segregation of Xg blood types—another X-linked marker—not a single case of recombination between G6PD and the centromere was observed. This is negative evidence. However, if G6PD was at the end of the X-chromosome long arm, it should not have been difficult to find evidence of meiotic recombination between this marker and the centromere in this special type of Klinefelter family (Filippi, G., and Siniscalco, M., *unpublished*).

Ruddle: Those data don't rule out the presence of G6PD either near the centromere or near the end of the chromosome.

Siniscalco: The data for the moment do favor the location of G6PD near the centromere. It would take too long to explain why. We are using the same approach utilized by Beadle and Emerson [Beadle, G., and Emerson, S., (1935): *Genetics 20*:192] in Drosophila and by Lindsley et al. [Lindsley, D., Fankhauser, G., Humphrey, R., (1956): *Genetics 41*:58] in the axolotl, for the very same purpose of mapping distances of given markers from their centromere.

Gerald: You could just as easily argue that the Klinefelter chromosome anomaly occurs because there is something wrong with the mitotic process and recombination.

Siniscalco: The same argument could be used.

Gerald: We simply don't have enough data to be critical at this time, I would say.

Littlefield: Does anybody have information where on the X-chromosome is the α-galactosidase locus, whether it is structural or regulatory?

Siniscalco: I think we have provided such data [Grzeschick, K., Grzeschick, A., Benoff, S., Romeo, G., Siniscalco, M., van Someren, H., Meera Khan, P., Westerveld, A., and Bootsma, D., (1972): *Nature New Biol.* *240*:48].

Littlefield: On linkage?

Siniscalco: No.

Bootsma: We are studying this locus in these translocation hybrids, and the first results indicate an independent segregation of α-galactosidase.

Watkins: I would like to go back to this question of reversion of A9 to HGPRT positive. I don't think Harris would mind if I give his explanation of this. The explanation is very simple, that all these cultures are contaminated with L cells. This explanation could be obviously demolished instantly if somebody who has observed revertants will get up and say there are no L cells growing in his laboratory.

Gerald: May it be noted that Dr. Evans stood up.

Chan: We also have HGPRT minus 3T6 cells. When we try to hybridize these cells, one out of every three clones of survivors is a revertant.

Basilico: If the suggestion of Pontecorvo is true, the reversion should be partial. In other words, they could be partial revertants that need some kind of initial stimulus in order to start dividing. This should imply that they have a low-plating efficiency. I also wonder about the possibility that this represents reversion of a regulatory type of mutation. Metabolic cooperation during hybridization experiments could induce the cells to make enzymes which they normally don't make. Does anybody have any evidence on these points whether A9 makes a nonfunctional enzyme or makes nothing?

Siniscalco: Dr. Shin has obtained evidence, using an antibody to mouse HGPRT, that there is some enzyme produced in at least one of the A9 lines. I would also like to point out that the frequency with which the HGPRT plus revertant mutates back to HGPRT minus is very much higher than that of the original L cells to HGPRT deficiency (Shin et al., 1973, cited above).

Somatic Cell Hybridization, edited by
Richard L. Davidson and Felix de la Cruz.
Raven Press, New York © 1974.

X-Autosome Translocations in Hybrid Cells

Park S. Gerald

The Children's Hospital Medical Center, 300 Longwood Avenue, Boston, Massachusetts 02115

I will present some data which we have obtained concerning a translocation between the X-chromosome and chromosome 19. (These studies were performed in collaboration with Drs. Gail Bruns and Virginia Monedjikova.) Figure 1 shows representations of these two chromosomes. I want you to note that my diagram of the banding pattern of the X-chromosome is quite different from Ruddle's. The breakpoints in our translocation are marked by arrows. (The breakpoint in the X-translocation studied by Ruddle occurred between the centromere and the proximal dark band on the long arm of the X.) In chromosome 19 the break is somewhere near the middle part of the long arm of chromosome 19.

Leucocytes from the patient carrying the X/19 translocation were hybridized with mouse (RAG) cells. The results are presented in Table 1. (The presence of HGPRT was inferred from survival in HAT.) Phosphoglycerate

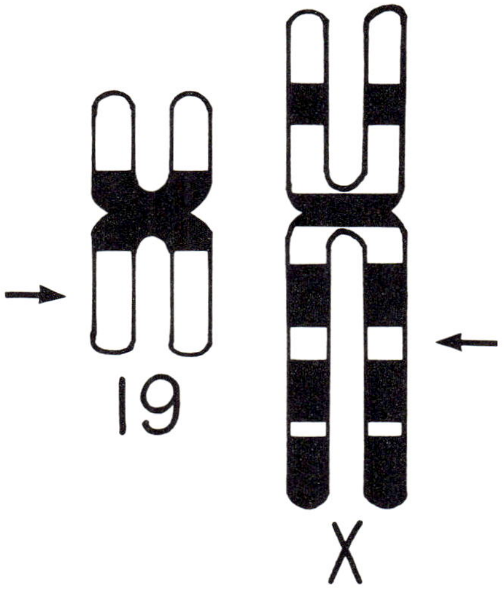

FIG. 1. Schematic representation of human chromosomes 1 and 19. The arrows indicate the sites of the breakpoints in the *t(Xq⁻; 19q⁺)* translocation.

TABLE 1. *RAG X WBC, t(Xq−; 19q+)*[a]

Clone	Chrom.	PHI	PGK	G6	HG
5X	1s	+	−	+	(+)
6	1s	+	−	+	(+)
11	1s	+	−	+	(+)
10	2s	+	−	+	(+)
15X	2s	+	−	+	(+)
17	2s	+	−	+	(+)
19	2s	+	−	+	(+)
1X		+	−	+	(+)
18X		+	−	+	(+)
9	1s	+	−	+	(+)
3X	1s	−	−	−	(+)
4	1s	−	−	−	(+)

[a] Breakpoint in X between first and second dark bands.

kinase (PGK) is always absent. This was totally unexpected. If PGK is near the centromere and proximal to the breakpoint, the translocation element containing PGK and the X-centromere should have been retained randomly, and some positive clones should have been observed. We were concerned that perhaps the breakpoint actually involved the PGK locus and suppressed its expression. However, fibroblasts from this patient are PGK positive. Unless there is something very peculiar, such as the white cells being totally unrepresentative of the fibroblasts, the results suggest nonrandom loss of a chromosome. It should be pointed out that phosphohexose isomerase (PHI

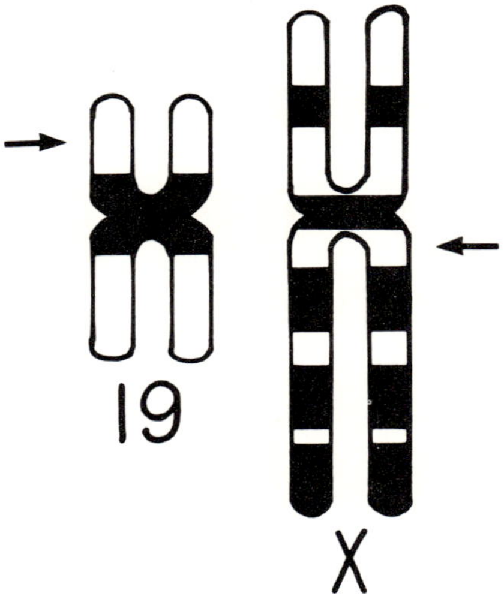

FIG. 2. Schematic representation of the breakpoint in the *t(Xq⁻; 19p⁺)* translocation.

or GPI), which has been assigned to chromosome 19 by Ruddle, is present in those clones expressing G6PD.

We have studied a second translocation between chromosome 19 and the X-chromosome (Fig. 2). The breakpoint in the X in this translocation, like that in the KOP translocation, is proximal to the first dark band. We believe that the breakpoint is on the short arm of chromosome 19. Table 2 shows that PGK, G6PD and HGPRT are retained in a majority of these clones.

TABLE 2. *RAG X WBC, t(Xq−; 19p+)*[a]

Clone	Chrom.	PHI	PGK	G6	HG
A4	1s	+	+	+	(+)
B1	1s	+	+	+	(+)
B2	1s	+	+	+	(+)
B5	1s	+	+	+	(+)
B7	1s	+	+	+	(+)
A3	2s	+	+	+	(+)
A8		+	+	+	(+)
A9		+	+	+	(+)
A10	1s	−	+	+	(+)
B9	1s	+	+	+	(+)
B10	1s	+	−	+	(+)
A2	1s	−	−	−	(+)

[a] Breakpoint in X between centromere and first dark band.

The results summarized in Table 2 are in accord with the data given by Ruddle and suggest that PGK, G6PD, and HGPRT are all on the long arm of the X. The data from Table 1 indicate that PGK is separable from the doublet G6PD-HGPRT. We believe that the arrangement is either centromere-PGK-G6PD-HGPRT or centromere-PGK-HGPRT-G6PD.

Somatic Cell Hybridization, edited by
Richard L. Davidson and Felix de la Cruz.
Raven Press, New York © 1974.

Linkage Studies on Adenylate Kinase

Jean Frézal

Unite de Recherches de Genetique Médicale (INSERM), Hopital des Enfants Malades, 149, rue de Sevres, Paris XV, France

I will discuss some work on adenylate kinase performed in my laboratory by Nguyen Van Cong, Regis Rebourcet, and colleagues (2). As previously shown by Brock, the activity and electrophoretic pattern of adenylate kinase varies from tissue to tissue (1). Usually adenylate kinase displays three bands, which are denoted 1, 2, and 3 (from the anode to the cathode). In red cells, as well as in muscle and fibroblasts, band 3 is the major component. Band 2 is less prominent and band 1 is faint. However, band 3 is very weak or absent in fresh lymphocytes, and is totally absent in cultured lymphocytes. Band 2 is the major component in lymphocytes. In liver and kidney, band 2 is strong while band 1 and 3 are of moderate and weak intensity, respectively. In man-mouse hybrids, the human band 1 overlaps one of the mouse bands, and human adenylate kinase is recognized in these hybrids by the presence of bands 2 and 3, which are clearly separated from the bands of mouse adenylate kinase.

The present observations were made on 42 primary clones (35 from fusions of mouse cells—LM(TK$^-$) clone 1D—with human fibroblasts and 7 from fusions of these mouse cells with human lymphocytes) and 30 subclones established from clones positive for adenylate kinase 2. Adenylate kinase band 3 was absent in all the hybrids, whereas adenylate kinase band 2 was present in 46 cases and absent in 26 cases. The question arose whether the adenylate kinase band 2 observed in the hybrids could be an altered band 3. If hemoglobin is added to the hybrid cell extract, band 2 does not overlap band 3 of red cells. It would therefore appear that, in the hybrids, bands 2 and 3 do segregate independently, suggesting that they are under the control of two different loci. This conclusion was strengthened when we found several hybrids which displayed both adenylate kinase bands 2 and 3 and one hybrid which was positive for band 3 and negative for band 2.

The well-known pattern of adenylate kinase in red cells can, in all likelihood, be attributed to the common AK_1 locus. Genetic variants at this locus are known and are characterized by a shift in the electrophoretic mobility of all three red cell adenylate kinase bands. Adenylate kinase band 2, in fibroblasts, lymphocytes, and perhaps liver and kidney is under the control of the AK_2 locus. I do not, however, understand why band 2 is completely absent when band 3 is present in the hybrids.

We have also correlated the expression of adenylate kinase in the hybrids with that of some 20 other enzymes and have found a strong correlation of AK_2 with peptidase C and PGM_1, as shown in Table 1.

TABLE 1. *Linkage study of human adenylate kinase and other markers in man-mouse hybrids*

				Hybrids Man-Mouse				
				Cl.1D-Fibroblast clone	sub-clone	Cl.1D-lymphocyte		
AK_1	AK_2	PGM_1	Pep C					
−	+	+	+	20	20	0	Association	
−	−	−	−	12	8	6	66	
−	(+)	(+)	−	1	0	0	Dissociation	Detection sensitivity
−	+	+	−	0	2	0		is in order
−	(±)	−	−	2	0	0	5	$AK \geqslant PGM_1 >$ Pep C
−	+	(−)	+	0	0	1	Dissociation	Human parent is het-
							1	erozygous for PGM_1
				35	30	7	72	

We have 6 discordant clones, including one hybrid, which is plus minus plus. We have interpreted 5 of these discordant clones to represent a dilution effect. The order of sensitivity of detection of these markers is $AK > PGM_1 >$ peptidase C. You can expect to have in some clones only very few chromosomes carrying AK_2, PGM_1, and peptidase C. Such clones would be scored plus plus minus or plus minus minus.

The hybrid which is plus minus plus was obtained from an individual heterozygous at the PGM_1 locus. The gene product of the second allele at the PGM_1 locus cannot be distinguished from one of the mouse PGM bands. Hence, if the chromosome carrying the human PGM_1 allele was lost in the hybrid, such a clone would be scored as PGM negative.

In summary, we have presented evidence for a second locus for adenylate kinase and have demonstrated synteny of this locus with peptidase C and PGM_1. The second adenylate kinase locus can therefore be assigned to chromosome 1.

REFERENCES

1. Brock, D. J. H. (1970): Adenylate kinase isoenzyme patterns in tissues of people of different phenotype. *Biochem. Genet. 4*:617.
2. Nguyen, Van Cong, Billardon, C., Rebourcet, R., Le Borgne de Kaouel, C., Picard, J. Y., Weil, D., and Frezal, J. (1972): Existence probable d'un second locus de Ak lié au groupe PGM_1-Peptidase C. *Ann. Genet. 15*:213.

DISCUSSION

Ruddle: Which AK is this, AK_2?

Frezal: Yes. The terminology is poor. In our system, AK_3 is the most cathodal of the three bands, and corresponds to red cell adenylate kinase.

Gerald: The three-banded pattern comes from what tissue?

Frezal: You see this pattern in fibroblasts. In the man-mouse hybrids, we observe strong activity at the AK_1 region, due to overlapping of the most anodal human band and one of the mouse bands.

Siniscalco: When you add hemoglobin, does the pattern become clearer?

Frezal: The hemoglobin slows down the migration of AK. If you compare the mobility of cellular extracts without hemoglobin with extracts of red cells, you see a difference in migration. But if you add hemoglobin to the cell extracts, the migration of the bands becomes comparable to those of red cells.

Siniscalco: Can you get better differentiation between mouse and human AK in that case?

Frezal: No. The same effect is observed with mouse adenylate kinase.

Shows: We also get these three bands in a different electrophoretic system. In several of our clones, we also see a segregation between AK_2 and AK_3, suggesting that independent loci code for these two forms. We haven't observed any linkage between AK_2 and PGM_1 or peptidase C as yet.

Bootsma: I will present two observations which Dr. Jongsma and Dr. Burgerhout have recently made in our institute which contribute to the mapping of chromosome A-1. In some Chinese hamster–human clones, a spontaneous breakage of chromosome A-1 occurred. In one such clone, positive for PGM_1 and 6 phosphogluconate dehydrogenase (6PGD) but negative for peptidase C (PEP-C) we found a chromosome having retained the short arm of chromosome A-1 and lost most of the long arm including the secondary constriction. The break point is probably in region 1*p*11 or 1*q*11. The centromere may be of Chinese hamster origin (see Discussion Fig. 1). We have analyzed 9 subclones, 8 of which have retained this marker chromosome and are 6PGD, PGM_1 positive, and PEP-C negative. One subclone has lost this chromosome and is negative for all three markers. Although we need to analyze additional clones, the data would suggest that PGM_1 and 6PGD are on the short arm of chromosome A-1.

The second observation was made on irradiated Chinese hamster–human hybrids. The original clone had retained the human X, chromosome A-1, and chromosomes 11 and 12 during many passages. This clone was irradiated with 600 R and 30 subclones were isolated. We were surprised to find the human X and chromosomes 1, 11, and 12 in these subclones. However, one of the subclones (HRB2) had lost 6PGD but retained PGM_1 and PEP-C. This subclone contains a chromosome which appears to have the long arm and part of the short arm of the human A-1 chromosome, having

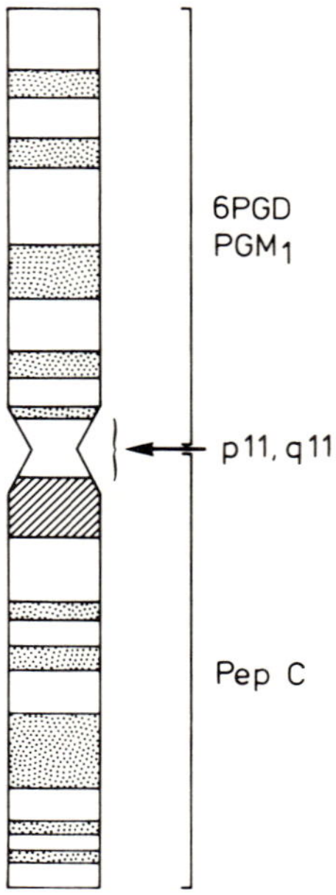

FIG. 1. Schematic representation of chromosome A-1, indicating break point found in a human–Chinese hamster somatic cell hybrid clone (see Ans Jongsma et al. *(in press): Human Genetik*).

lost the distal part of the short arm (see Discussion Fig. 2). This observation suggests that 6PGD may be on the distal part of the short arm of chromosome A-1.

Ruddle: The linkage map indicates 6PGD might be 116 map units from the centromere. This might still be consistent with the loss of the material from the q arm, deleting PEP-C. There would still possibly be enough room for all of these markers to fit into the remaining piece between the break point and the centromere. We have observed one clone that has a similar deletion of the distal part of the short arm. However that clone retains all three of the markers, or at least PEP-C.

Gerald: You mean that the data for chromosome 1 is as confusing as for the X?

Bootsma: The first deletion discussed was a deletion of the long arm of chromosome 1, while the second deletion was a deletion of the short arm (clone HRB2).

Gerald: But you claim, do you not, that 6PGD and PGM are most likely on the short arm?

Bootsma: That is the most reasonable explanation.

Renwick: I should like to say that I don't think there are any data from family studies that are irreconcilable with a map position of 6PGD near the

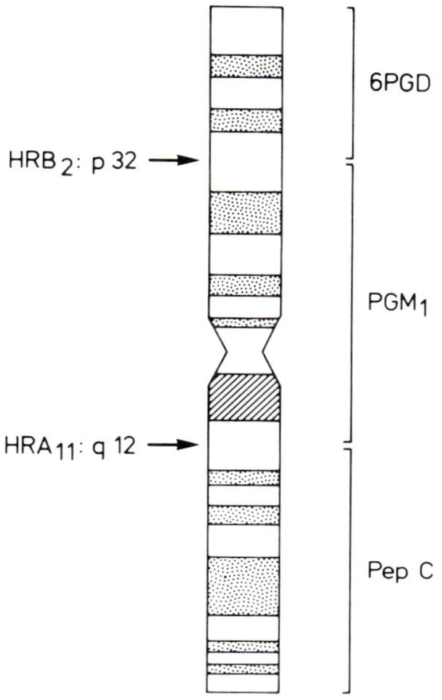

FIG. 2. Schematic representation of chromosome A-1. Two different break points introduced by X-irradiation in a human–Chinese hamster somatic cell hybrid are indicated. Clone HRB2 is presented in the discussion. Clone HRA11, missing the distal part of the long arm and PEP-C, has not been discussed. (See Burgerhout et al. *(in press): Human Genetik*).

end of the map of the short arm. The map of chromosome 1 would still be roughly the same as had been suggested except that everything would be on the short arm instead of the long arm. But it is still very tentative.

The data from translocations in family studies are weak and, in part, conflicting. Some of the unpublished Edinburgh-Galton data are slightly in favor of the linkage group, PGM_1-6PGD being on the short arm.

Gerald: Could you say whether or not localization of Duffy and amylase by recombinational analysis with the heterochromatic region is sufficiently precise to assign these to the short or long arm?

Renwick: I would say that those loci map near the centromere, but we do not know on which side of the centromere.

Evans: In observations on around 150 people, we have found two inversions involving the centromere zone of chromosome number 1. In rather larger samples of the population, we find that some 2% of individuals have inversions involving the centromeric region of chromosome 9 and a proportion show similar pericentric inversions in chromosome 8. Many of these inversions are only revealed through using banding techniques and these methods clearly show that, for a given chromosome, there may be a variety of different sized inversions present within the population. If we are allowed to generalize from these data, it would seem likely that pericentric inversions may be very frequent phenomena in human populations and this is an important factor which must be taken into account when discussing the location, relative order, and map distances of genes situated in, or close to, these centromeric regions.

Renwick: Among other things, it would tend to make the crossing over in that region appear less than expected from the actual physical lengths.

Minna: Does anybody have evidence that either X irradiation or some chemical treatment can be associated with recombination within a chromosome linkage group of an isozyme or gene product rather than changes in chromosome structure alone?

Coon: John Minna and I have looked at about 300 independently derived hybrids between mouse or rat and human cell lines. We have observed segregation of the rodent chromosomes in 90% of the hybrids. This observation may have some value for those who want to determine which chromosomes will be lost. We can find this phenomenon in hybrids between mouse, rat, and even Chinese hamster cells and a number of human cell lines. Most of our results have been with either VA-2 or D98 AH-2 and mouse cells freshly liberated from the central nervous system, heart, liver, bone marrow, spleen, or thymus.

We have very preliminary information that this principle that the chromosomes of freshly liberated cells are preferentially at risk may apply to other species combinations besides human-mouse and human-rat.

Somatic Cell Hybridization, edited by
Richard L. Davidson and Felix de la Cruz.
Raven Press, New York

The Fate of the Mitochondrial DNAs in Mouse-Human and Rat-Human Hybrid Cells

Hayden G. Coon, Ivan Horak,* and Igor B. Dawid*

*Laboratory of Cell Biology, National Cancer Institute, National Institutes of Health, Bethesda, Maryland 20014 and *The Carnegie Institution of Washington, Department of Embryology, Baltimore, Maryland*

We have been interested in studying the inheritance of mitochondrial DNA in hybrid cells. Our studies have been performed with mouse-human and rat-human hybrids which showed chromosome segregation patterns in about 90% of independently arising hybrids opposite to that normally observed, i.e., in these hybrids the rodent chromosomes were the ones most often preferentially lost as the hybrid strains were propagated (3).

We have distinguished the mitochondrial DNA (mtDNA) of the different species by detecting species specific nucleotide sequences using a molecular hybridization assay developed by Dawid (4). To calibrate the assay, known quantities of purified mtDNA of the two species are attached to Millipore filters and then hybridized with complementary RNAs (cRNAs) synthesized with *E. coli* RNA-polymerase using highly purified mtDNA of the two species as template. The cRNA of one species is labeled with tritium and that of the other species with ^{32}P. The isotope ratio is then plotted against the percentage of mtDNA of each species in the mixture on the filter. The relative amounts of each species' mtDNA in the hybrids are then determined from the isotope ratio by reference to the calibration curve. A calibration curve is run in the same vial with each sample of unknowns acting as a control for possible variability among different batches of cRNAs produced by the *E. coli* RNA-polymerase.

We, as well as a number of other investigators, found that nuclear DNA does not hybridize significantly with cRNA made from purified mtDNA. Therefore, the same calibration curves are obtained whether the hybridization is done with purified mtDNA or with whole cell DNA on the filters. Since the nuclear DNA does not interfere with the assay for mtDNA, it is possible to use a small number of cells for the analysis. For example, $2-4 \times 10^6$ hybrid cells provides sufficient DNA for the detection of mtDNA of one species in amounts as small as 1% of the total mtDNA.

We have used the same type of calibration curve to determine the relative amounts of nuclear DNA of the two species in the hybrid cell populations. It is only the repeated sequences of DNA that are hybridized under these conditions, but there is usually a good correspondence between the number of chromosomes of one species in the hybrid and the percent of nuclear DNA determined by hybridization for that species. The hybridization assay agrees to within about 10% with the actual count of chromosomes which has been determined for samples of several mouse-human hybrid strains (3).

The results of the measurement of the percent of rat and human mtDNA and nuclear DNA in a large series of hybrids is shown in Fig. 1. Those hybrids that have a higher percentage of human chromosomes (i.e., are tending to lose rat chromosomes) also tend to have a higher percentage of human mtDNA (and have tended to lose rat mtDNA). Similarly those hybrids that tend to lose human chromosomes also lose human mtDNA.

Results of this kind alone cannot establish that the mtDNA of both species exists within the same individual cells of the hybrid population. The hybrid population could have segregated into two subpopulations, one with

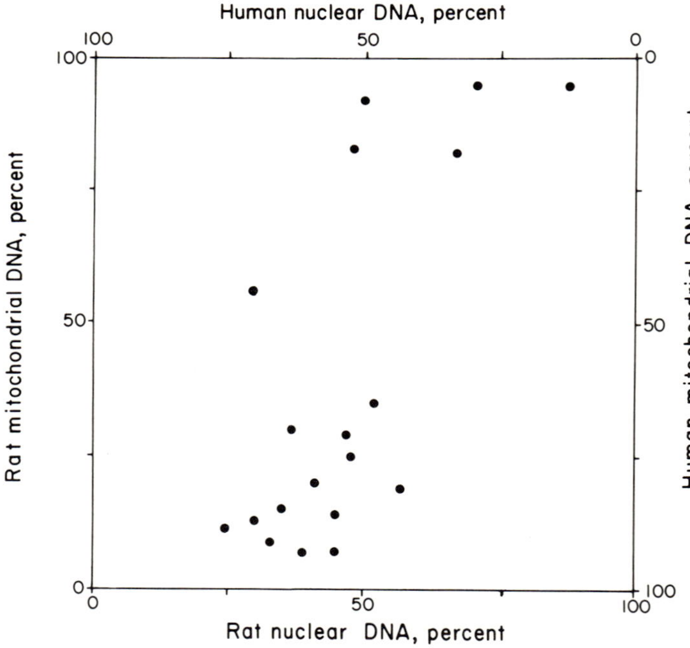

FIG. 1. The mitochondrial (mt) and nuclear (n) DNA composition is shown for 19 independently derived rat-human hybrid cell strains determined approximately 60 population doublings after fusion. Each point is plotted according to that strain's rat nDNA and rat mtDNA (determined in separate cRNA/DNA hybridization experiments—see text) and is expressed as a percent (the human DNA content is 100% minus the rat DNA content). The plot shows that after 60 doublings, the hybrid populations have tended to cluster into two groups: the larger group is losing both rat n and mtDNA, whereas a smaller group is losing human n and mtDNA (high in rat n and mtDNA).

mtDNA of one species and a second subpopulation with mtDNA of the second species.

To resolve this point, the original hybrids were subcloned after 60 to 80 population doublings in culture, and the mitochrondrial and nuclear DNA of populations grown from the subclones were analyzed (Fig. 2). In every case but one (not illustrated), all of the subclones had a mixed population of mtDNA; in that one case, all the clones were of one type, indicating that mtDNA of the other species had been lost altogether at the time of cloning. These results have proven that in many of the hybrids the mtDNA of the two species replicates in the same cytoplasm for extended periods of time.

Similar results were obtained with mouse-human hybrids (see Fig. 3). In these hybrids, chromosome segregation is more extreme. Those hybrids high in mouse nuclear DNA contain practically exclusively mouse mtDNA.

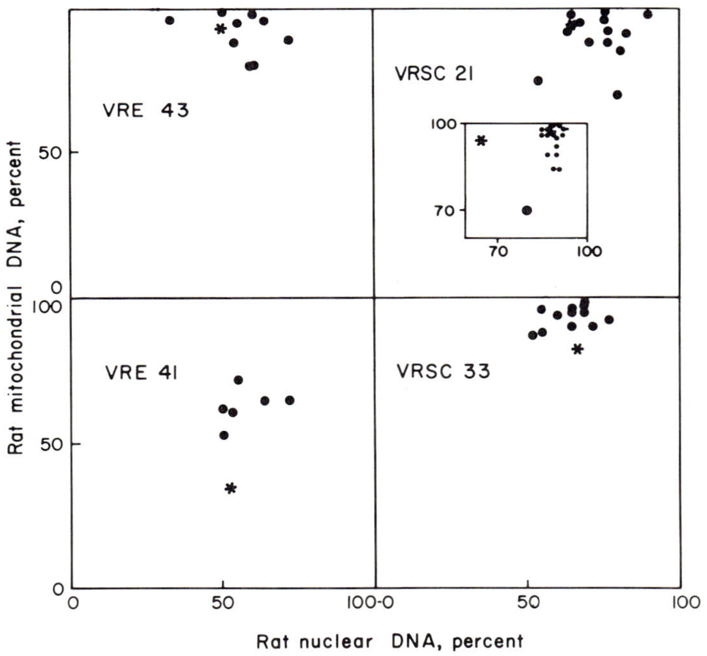

FIG. 2. Four rat-human hybrid cell strains were subcloned and the n and mtDNA determined and plotted for the subclonal populations as in Fig. 1. Each of the subclones of the four strains shows presence of both the parental n and mtDNAs. The star represents the values for the parental population at the time of subcloning. The inset on VRSC 21 shows a series of sub-subclones. The original VRSC 21 colony is shown as a star in the inset and the subclone from which the sub-subclones were derived is shown as a large circle. Since all of the subclones and sub-subclones of these hybrid populations remain hybrid for the mtDNAs it is possible to conclude that mtDNAs of both species are propagated for many generations (over 100 for the sub-subclones) in the individual cells of the hybrid populations.

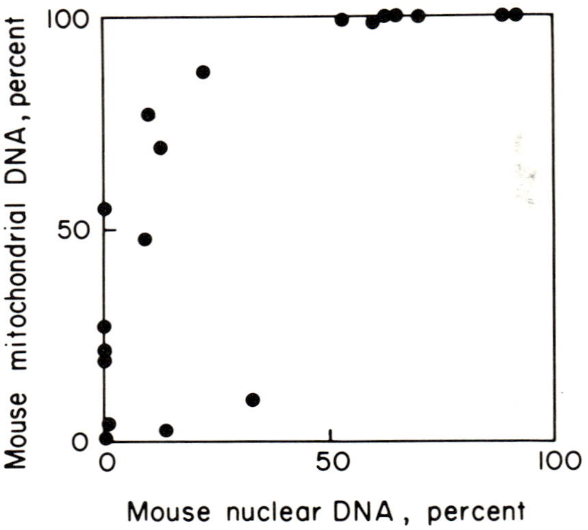

FIG. 3. The n and mtDNA composition is shown for 19 independently derived mouse-human hybrid cells strains determined 40 to 60 population doublings after fusion. The points were obtained and plotted as described under Fig. 1. Chromosome segregation, like the n and mtDNA segregation shown, is more extreme in our series of mouse-human hybrid strains than in our rat-human hybrids (both series are unusual in that they frequently lose rodent chromosomes). Mouse mtDNA seems to persist in mouse-human hybrids longer and with less mouse nDNA support than rat mtDNA does in the rat-human hybrids. When hybrids with 0 (≤3%) mouse nDNA were retested about 25 doublings later, mouse mtDNA was also undetectable (≤1%) in the populations.

Such hybrids are similar to those studied by Clayton et al. (2) and Attardi and Attardi (1). Most of this series of hybrids, like the rat-human hybrids, were segregating rodent chromosomes. Some hybrids that have lost all detectable mouse nuclear DNA (≤3%) have nevertheless retained significant amounts of mouse mtDNA. When such hybrids were tested about 20 population doublings later, however, the amount of mouse mtDNA had also dropped to undetectable levels. This observation strengthens the conclusion drawn from the analyses that, if the chromosomes of one species are lost from the hybrids, then the mtDNA of that species is also lost.

We find these hybrids most intriguing and are interested in studying the relationships between the nuclear and mitochondrial genomes. We are also looking for the possibility of recombination between the mitochondrial DNAs of the two species in the hybrids. Our preliminary results suggest that recombination may indeed be occurring. Using density gradient centrifugation to separate the parental mtDNA species in human-mouse hybrids, we have detected sequences of mouse mtDNA in the peak of human mtDNA and vice versa. Both the amount of cRNA bound and the calculated density of these "recombinant peaks" is consistent with the idea that 10 to 20% of the nucleotide sequences in the peak at or close to human density are of mouse origin. We have not yet established whether these sequences are covalently bound to the mtDNA of the other species, but that seems to be the most likely interpretation of our gradients (5).

REFERENCES

1. Attardi, B., and Attardi, G. (1972): Fate of mitochondrial DNA in human-mouse somatic cell hybrids. *Proc. Nat. Acad. Sci. 69*:129.
2. Clayton, D., Teplitz, R., Nabholz, M., Dovey, H., and Bodmer, W. (1971): Mitochondrial DNA of human-mouse cell hybrids. *Nature 234*:560.
3. Coon, H., Horak, I., and Dawid, I. (1973): Propagation of both parental mitochondrial DNA's in rat-human and mouse-human hybrid cells. *J. Mol. Biol. 81*:285.
4. Dawid, I. (1972): Evolution of mitochondrial DNA sequences in *Xenopus. Devel. Biol. 29*:139.
5. Horak, I., Coon, H., and Dawid, I. (1974): Interspecific recombination of mitochondrial DNA molecules in hybrid somatic cells. *Proc. Nat. Acad. Sci.* May.

DISCUSSION

Pontecorvo: How did you make the hybrids that lost rodent chromosomes?

Coon: These hybrids were all isolated from crosses between an established human cell line, VA-2 (SV40 transformed), and freshly liberated cells from embryonic or adult mouse or rat tissues. Generally, the nervous system was used as the source of cells, but we do not believe that there is any difference using cells from the nervous system, bone marrow, thymus, spleen, liver, or heart.

Pontecorvo: What selection did you use?

Coon: The SV40 transformed human line was obtained from Mary Weiss. It is 8-azaguanine resistant and is selected against in HAT medium.

Graham: Was the chromosome complement of the hybrid that had a lot of mouse mitochondrial DNA but no detectable mouse nuclear DNA analyzed?

Coon: Yes, it was. After analyzing about 100 metaphases, we can say that about one cell in three has a recognizable mouse marker chromosome. VA-2 has a complicated karyotype with a number of rearrangement products that are not stable. It is, therefore, necessary to analyze the chromosomes by quinacrine fluorescence.

Migeon: Hadn't Weiss fused VA-2 cells with diploid rat cells and obtained hybrids that lost the human chromosomes?

Weiss: Ephrussi had crossed VA-2 cells with diploid mouse fibroblasts from secondary cultures. These hybrids did lose human chromosomes [Weiss, M., Ephrussi, B., and Scaletta, L. (1968): *Proc. Nat. Acad. Sci. 59*:1132].

Coon: I should add if we had used secondary diploid cultures, we also would see the segregation of human chromosomes. However, we are using rodent cells freshly liberated from tissues.

Somatic Cell Hybridization, edited by
Richard L. Davidson and Felix de la Cruz.
Raven Press, New York © 1974.

Induced Chromosome Elimination in Hybrid Cells

G. Pontecorvo

Imperial Cancer Research Fund, Lincoln's Inn Fields, London WC2, England

I have been continuing to investigate the elimination of chromosomes from hybrids produced by crossing cells irradiated (X- or gamma rays) immediately prior to fusion with nonirradiated cells (1). The system chosen is one in which the chromosomes of the 2 parent species could be distinguished and in which no substantial loss of the chromosomes of either parent from the hybrids occurs without irradiation. One of the parental lines was a TK-deficient mouse 3T3 line, with a mode of 66 acrocentric chromosomes and no biarmed chromosomes (kindly supplied by Dr. Basilico). The other parental line was an HGPRT-deficient Chinese hamster line, with 22 chromosomes, all distinguishable from the mouse chromosomes (kindly supplied by Dr. Westerveld). Hybrid clones between these two lines have almost exactly the sum of the chromosome complements of the parental cells (see Fig. 1). Chromosome loss from these hybrids is minimal over at least the first two months of subculture in HAT.

In a series of experiments either the Chinese hamster cells or the mouse cells were irradiated just prior to fusion with unirradiated cells of the other species, and the hybrids were isolated as clones within 10 to 20 days and maintained in HAT. The hybrid clones showed extensive loss of the chromosomes of the irradiated parent. A karyotype of one such hybrid is shown in Fig. 2. In this case there are only 11 hamster chromosomes left, a loss of 11 chromosomes. The present stage of the work is summarized in Table 1, for experiments with irradiation doses up to 1500 rads. The loss of chromosomes was found to be dependent upon the irradiation dose. For example, when the mouse cells were irradiated at 600 rads, there was a mean loss of 3.2 mouse chromosomes from the hybrids. At 1000 rads the average loss of mouse chromosomes was 11.3. At 1500 rads the average loss was 25.7. Similar results were obtained when the hamster cells were irradiated prior to fusion. At 600 rads, an average of 2.7 hamster chromosomes were lost from the hybrids, at 1000 rads 5.2 chromosomes, and at 1500 rads 10.9 chromosomes. Control hybrids in which neither parent cell was irradiated showed no substantial chromosome loss. It is clear that irradiation before fusion can cause the loss of many chromosomes of the irradiated parent from the hybrids.

Not surprisingly, after irradiation of one of the parental cells, fewer hy-

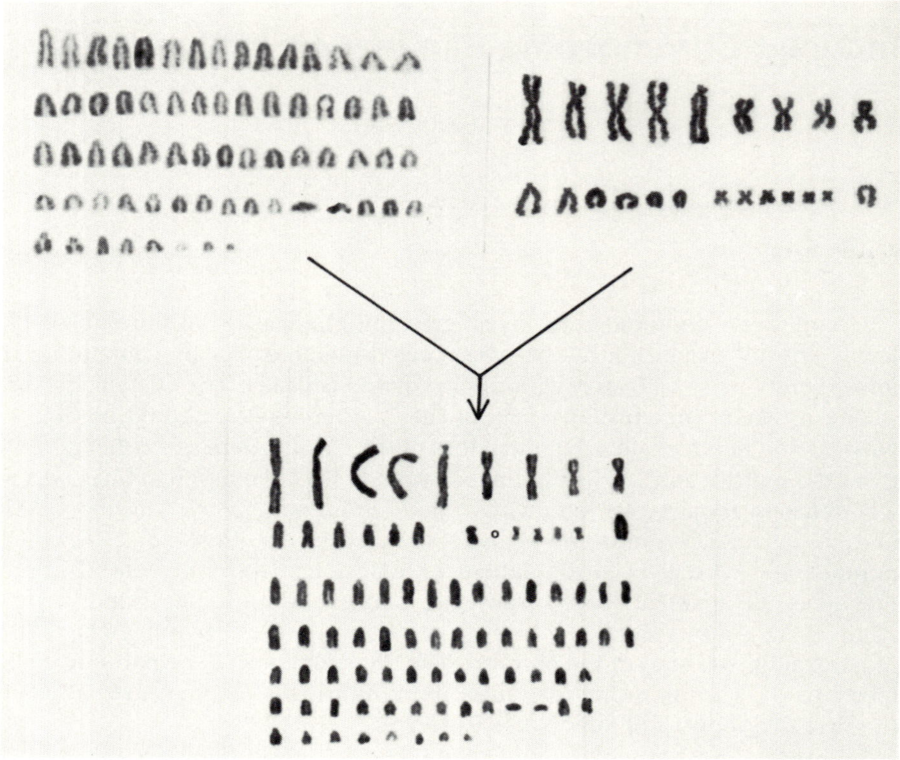

FIG. 1. Top left: karyogram of the mouse line. Top right: karyogram of the hamster line. Bottom: typical karyogram of a clone of hybrids between nonirradiated cells.

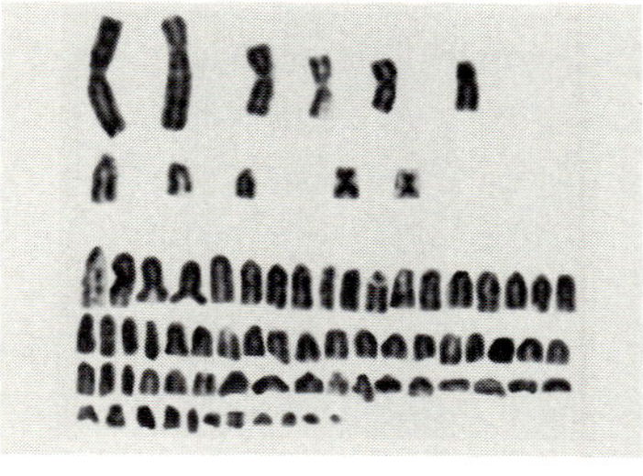

FIG. 2. Karyogram of a hybrid clone between irradiated hamster cells and nonirradiated mouse cells.

TABLE 1. *Loss of chromosomes of the irradiated parent*

Dose rads	Mouse cells irradiated: Clones no.	Mouse chromosomes lost: Mean	Mouse chromosomes lost: Range	Hamster cells irradiated: Clones no.	Hamster chromosomes lost: Mean	Hamster chromosomes lost: Range
0	21[a]	2.7			0.7	
600	6	3.2	−3:9	8	2.7	2:10
1000	13	11.3	4:25	7	5.2	3:11
1500[a]	13	25.7	12:51	11	10.9	7:17

[a] Provisional data, analysis in progress.

brids are formed. However, if the rate of hybrid formation is compared to the survival rate of the irradiated parental cells, it is apparent that there has been an enormous amount of rescue. Cells that are rendered inviable by irradiation are rescued by fusion with nonirradiated cells. This, of course, is the predicted result. For example, at 1500 rads the surviving fraction of irradiated hamster cells is reduced to 0.08%. The number of hybrid clones obtained when these irradiated cells are hybridized with nonirradiated mouse cells is about 2% of the number obtained when nonirradiated hamster cells are hybridized with mouse cells. Since the decrease in viability (over 1000-fold) is so much greater than the decrease in hybridization frequency (50-fold) it is clear that there has been significant rescue. Similar results have been obtained by Dr. Morgan Harris in hamster-hamster hybrids.

I have also obtained some information on the number of chromosome complements in these hybrids. Some of the hybrids (with either parent irradiated) contain a 2 S chromosome complement of the nonirradiated parent.

Hybrids with a 2 S complement of one or the other parent are also observed occasionally in control crosses. Significantly, in the hybrids with one irradiated parent, the 2 S contribution was always from the nonirradiated parent. This could be the consequence of fusion with tetraploid cells, which are always present in the populations in a small proportion, or of multiple fusion. However, the conditions of fusion were such as to minimize the possibility of multiple fusion. Furthermore, to account for the frequency of hybrids with a 2 S complement by fusion with tetraploid cells the pre-existing tetraploid cells would have had to fuse at a much higher rate than the diploid cells. Finally, the 2 S complement is either from the mouse or the hamster cell, depending on which parental cell was not irradiated. It is well known that following irradiation, there is a delay in chromosome replication in irradiated nuclei. My hypothesis is that this delay is carried over in the heterokaryons. In heterokaryons with one irradiated and one nonirradiated nucleus, in some cases, the nonirradiated nucleus has time to go through an additional division before fusion with the other nucleus.

It is interesting that when mouse cells are given high doses of irradiation the number of mouse chromosomes in the hybrids can be greatly reduced, as a maximum, in the range of irradiation used, from 66 chromosomes down to about 10. The same results are not obtained with heavily irradiated hamster cells. The hamster cells have 10 large metacentric chromosomes. These large chromosomes are less frequently found to be lost in the surviving hybrid clones. This is an unexpected result considering the mode of action of irradiation: in rough approximation, the larger a chromosome the greater its chance of undergoing breakage and successive elimination. Out of 26 clones from irradiated hamster cells, not one has fewer than five large metacentric hamster chromosomes. It would appear that there is very strong selective pressure to retain some of the large hamster chromosomes in the hybrids, even though the hybrids have the complete mouse genetic information. The results suggest that, even in mouse-hamster hybrids in which there is no substantial spontaneous preferential chromosome loss, there is something that acts to conserve part of the chromosomes of one species more than those of the other. Clearly, this is unlikely to be a simple matter of a full set of genetic information. It may be important to look at the interactions between two or more chromosomes and between centromeres and the spindle, for example.

Finally, I have tested this technique on man-mouse hybrids. If the mouse cells are irradiated before fusion with human diploid cells, the number of hybrids formed is greatly reduced. However, those hybrids that do grow segregate human chromosomes, as do hybrids with nonirradiated cells.

REFERENCES

1. Pontecorvo, G. (1971): Induction of directional chromosomal elimination in somatic cell hybrids. *Nature 230*:367.

DISCUSSION

Tashjian: The rescue experiments described by Pontecorvo are intriguing to me because of the observations we have made concerning the extraordinary degree of resistance to X-irradiation in hybrid cells [Little, J. B., Richardson, U. I., Tashjian, A. (1972): *Proc. Nat. Acad. Sci. 69*:1363]. We have seen this phenomenon with rat-mouse hybrid strains. We are far from understanding the mechanism; nevertheless, it is interesting that after hybridization you see rescue of cells that would have died unless they had been hybridized.

Thompson: Dr. Pontecorvo, how long after fusion did you characterize the hybrids?

Pontecorvo: Between 1 and 3 months after fusion.

Gerald: How much chromosome damage appears as a result of irradiation?

Pontecorvo: The results are very informative. If you look at the hybrid clones as soon as you can (2 to 3 weeks after fusion), you find many cells with chromosome aberrations as expected. However, if you look later, the retained chromosomes appear normal. In fact, there is very little of what I expected would make the chromosomal analysis difficult. There is little joining of two mouse chromosomes to form new metacentrics or cleavage of metacentric hamster chromosomes to form telocentrics. It seems that chromosomes are still being eliminated by some cells of a clone 20 days after irradiation, while the other cells have already "cleansed" themselves, no doubt by the usual breakage-reunion-breakage process.

Watkins: Have you ever tried irradiating both parents to see if there is some kind of mutual rescue?

Pontecorvo: I have wanted to do that but have not done it yet.

Siniscalco: Did you try to irradiate human cells and then fuse them with mouse cells?

Pontecorvo: Yes. It does not help very much.

Somatic Cell Hybridization, edited by
Richard L. Davidson and Felix de la Cruz.
Raven Press, New York © 1974.

Viruses, Immune Functions, and Antigenic Determinants in Heterokaryons and Hybrids*

Hilary Koprowski and Barbara Knowles

The Wistar Institute of Anatomy and Biology, 36th Street and Spruce, Philadelphia, Pennsylvania 19104

VIRUS RESCUE

The term "rescue" is explained in almost half a column in *Webster's New International Dictionary*. We shall use it in the sense of "to liberate from actual restraint," and a later discussion on the mechanisms of rescue will give you a more precise description based on experimental data presently available.

Prior to the introduction of the fusion technique, attempts to rescue oncogenic viruses ended in failure. The successful rescue of infectious virus through the fusion of SV40-transformed cells with permissive (i.e., capable of synthesis of infectious virus) African green monkey kidney cells (AGMK) (82) established a pattern for the rescue operation. SV40 could almost always be rescued (Table 1) from cells of SV40-transformed, nonpermissive (i.e., incapable of synthesis of infectious virus) mouse lines or from the progeny of numerous clones of these lines (43). Slightly different results were obtained after fusion of SV40-transformed, nonpermissive hamster cells with AGMK cells. Infectious virus could be rescued from some of the hamster lines (46) but cells of at least three of the lines failed to yield infectious virus after fusion (37). These latter findings may have to be revised in the light of observations that, following exposure to hamster syncytiogenic virus (38), SV40 could be rescued from at least five of the hitherto nonrescuable, transformed hamster tumor cells (39).

Defective SV40 seems to be involved in the transformation of permissive cells; on fusion with AGMK cells these transformed cells do yield a virus particle that causes microplaques in AGMK cells (34). These microplaques do not increase in size on long-term incubation of the infected plates. Transfer of tissue-culture fluid containing this presumptive defective virus induced T-antigen formation in a small percentage of both permissive and nonpermissive cells. Cesium chloride density gradient ultracentrifugation

* This review covers material published up to November, 1973; some pertinent references may not have been available to the authors.

TABLE 1. *Rescue of virus from transformed cells by fusion with permissive cells*

Virus group	Virus	Cells	Rescue of virus after fusion or hybridization
Papova	SV40	Transformed mouse cells	Rescue of SV40
		Transformed hamster cells	Rescue of either SV40 or defective SV40
		Transformed human or AGMK	Rescue of defective SV40
Adeno	Polyoma	BHK hamster transformed cells	None
		Rat embryo transformed cells	100 × increase in "spontaneous activation"
	Adenovirus 12	Transformed hamster cells	None
	SV40-adeno hybrid	Transformed AGMK	None
Oncornaviruses	MSV and RLV	Mouse cells shedding RLV	S–L+ S+L+ S+++L+
Herpes	EBV	Human transformed X human	None until induction with IUDR

AGMK = African green monkey kidney cells; RLV = Rauscher leukemia virus; MSV = murine sarcoma virus; S+ = sarcoma production; L+ = leukemia production; EBV = Epstein–Barr virus; IUDR = iododeoxyuridine.

of this rescued virus revealed a virus particle with a density 1.325 in contrast to the virus rescued from a SV40-transformed mouse cell whose density is 1.34. Mouse cells transformed by this microplaque virus on fusion with permissive cells again yield only microplaque-causing virus (36). Other investigators have shown that the plaque morphology of virus rescued by fusion from transformed nonpermissive mouse cells is identical to that of the original transforming virus (86).

Fusion of polyoma-transformed hamster cells (which are nonpermissive for virus) with permissive cells in general did not result in the rescue of the virus (17). When hamster cells were transformed by a temperature-sensitive mutant of polyoma virus at a permissive temperature and then maintained at a nonpermissive temperature it was possible to recover the virus after fusion with permissive cells. When the transformed hamster cells were maintained at permissive temperature, however, recovery of virus was not possible. These seemingly paradoxical results were explained (17) by the hypothesis that if in nonpermissive cells genetic functions of the virus are not completely expressed, the virus can still be recovered after fusion with permissive cells. In rat cells transformed by polyoma virus, spontaneous activation of viral synthesis may occur; when these cells are fused with permissive mouse cells, synthesis of the virus is increased in the heterokaryocytes (22).

Adenovirus cannot be rescued from transformed nonpermissive hamster cells after fusion with permissive cells (8). This may be due to the presence in the transformed cells of a factor restricting replication of adenovirus. The SV40-adeno hybrid virus can transform AGMK cells; however, up to now, defective SV40 could not be rescued from such cells after fusion with permissive AGMK cells (47). It seems that, at present, the rescue of SV40 from transformed cells is a phenomenon for which no parallel can be found in rescue studies with other tumor viruses, such as polyoma or adeno, which remain essentially in the "nonrescuable" category.

Fusion of Burkitt lymphoma–derived lines containing the latent Epstein-Barr virus (EBV) genome with non-EBV-transformed human cells or with another human non-EBV producer line does not result in production of EBV virus or viral antigens. Treatment of these cells with iododeoxyuridine results in the synthesis of EBV specified early antigens indicating that the genome is still present although not expressed in these hybrid cells (66).

Marmoset cultures infected with herpes saimiri virus were fused with mouse cells and the hybrid cells, in which loss of marmoset chromosomes was observed, were monitored for virus production (70). Some virus could be rescued from the original hybrid population on cocultivation with susceptible monkey kidney cells. Nonclonal populations gradually became nonrescuable presumably due to loss of the chromosome necessary for the virus integration or replication.

Although the mechanism of oncorna virus transformation of cells seems to be more complex than that of the oncogenic DNA viruses, rescue of virus also occurs on induction of the integrated leukemia genome by chemical means (88) or after introduction of a leukemia virus into a cell containing a sarcoma genome by infection, cocultivation or cell fusion. In this latter case, the defective sarcoma genome is rescued by the concurrent replication of the leukemia virus resulting in the production of infectious virus of both types. For example, hamster cells carrying defective murine sarcoma virus were hybridized with mouse cells shedding Rauscher leukemia virus. Each of eight hybrid clones examined showed the presence of a group-specific antigen of murine leukemia viruses and all produced infectious murine leukemia virus (Table 1). Seven of the eight hybrid cultures also produced sarcoma virus, but in contrast to mouse-cell carriers of both sarcoma and leukemia viruses, four of the hybrid clones produced an excess of sarcoma virus. These results seem to show that genomes of both viruses, sarcoma and leukemia, were present in the hybrid cells and that hybrid cells were capable of replicating both agents (58).

The successful rescue of some of the tumor viruses after cell fusion opened the possibility of utilizing the same system to isolate viruses from cells of patients with diseases of a suspected viral etiology. For obvious reasons, these studies were first aimed at the so-called slow virus diseases characterized by chronic involvement of the central nervous system (CNS) since routine methods of virus isolation in these cases had met with failure after failure.

Subacute sclerosing panencephalitis (SSPE) (Table 2) is a disease of the CNS characterized by a protracted course suggestive of slow virus infection. Brain cells obtained from explants of tissue fragments obtained at either biopsy or autopsy were maintained in culture for many cell transfers, the cells forming syncytia, showing the presence of myxovirus nucleocapsids by electron microscopy and staining for the presence of measles-like antigen in an immunofluorescent test. When these cells were fused with AGMK cells (51), there was no difficulty in isolating a virus related to the measles-distemper-rinderpest group. The rescued virus was found to cause encephalitis in ferrets (42). What is perhaps more interesting is the fact that another

TABLE 2. *Viruses in human chronic CNS diseases; detection of virus by fusion of human brain cells with indicator cells*

Disease	Fusion of brain cells with:	Type of virus isolated
Subacute sclerosing panencephalitis	AGMK	M-D-R group and papova-like virions[a]
Progressive multifocal leukoencephalopathy	AGMK	SV40
Measles encephalitis	AGMK	Measles
Multiple sclerosis	AGMK	Parainfluenza 1

M-D-R = Measles-Distemper-Rinderpest group.
AGMK = African green monkey kidney cells.
[a] As yet unidentified biologically.

virus was also found in the heterokaryocytes formed by the fusion between brain cells with AGMK cells (50). These papova-like virions were observed in the cytoplasm of the SSPE brain cells and in the progeny of AGMK cells fused with brain cells in which the paramyxovirions were also present. Except for ultrastructural observations and possible histochemical evidence of the presence of DNA inclusion bodies in the cytoplasm (92), no biological role for the papova-like virions can be assigned at present as attempts to isolate them in infectious form have been futile (52).

Cell fusion between human brain cells derived from patients suffering from progressive multifocal leukoencephalopathy (PML) and AGMK cells led to the isolation of infectious papova virus. These patients had no cancer, but the isolated virus was found to be similar to, or identical with, the SV40 (96) and when transmitted to hamsters caused tumor formation and not a demyelinating disease as in man.

Not surprisingly, fusion of brain cells growing out from explants of a case of chronic measles encephalitis yielded measles virus (91).

Isolation of parainfluenza 1 virus from brain cells originally obtained from two cases of multiple sclerosis (MS) (90) and fused with AGMK cells followed the same pattern as in the cases described above. There was only one curious twist to the latter story. MS cells were fused with AGMK cells either in the presence of inactivated Sendai virus or lysolecithin. Heterokaryocyte cultures formed as the result of fusion of parental cells in the presence of Sendai virus showed no lesions whatsoever, and no infectious virus was isolated from these cultures. In contrast, heterokaryocytes formed as the result of exposure of the parental cells to lysolecithin showed nuclear polymorphism, occasionally formed giant cells, and upon passage had a much higher mitotic rate than control cells. These cells showed positive hemadsorption with either guinea pig or human red blood cells and the presence of the nucleocapsids of paramyxovirus in the cytoplasm and nucleus (90). The absence of lesions in the Sendai-induced fusion cultures may perhaps be explained as interference by inactivated Sendai with the replication of another member of the same group of viruses. The transient inhibitory action of live or β-propiolactone-inactivated Sendai virus on polyoma virus synthesis in permissive mouse cells has been described (77). In this case, it was postulated (78) that Sendai virus RNA, apparently localized in the nucleus of the infected mouse cell, interfered with the encapsidation of DNA into mature polyoma virus particles. UV-irradiated Sendai virus, which retained its hemagglutinating activity, did not interfere with polyoma virus synthesis, probably because the biological activity of its RNA component was rendered nonfunctional.

FACTORS AFFECTING RESCUE OPERATION

In general, factors affecting the efficacy of cell fusion and hybridization have an effect on virus rescue (12). Of these factors, the environmental pH of the medium in which the cells grow at the time of fusion and while the heterokaryons are maintained seems to have a profound effect. As shown in Table 3, a definite increase in the percentage of multinucleated cells and heterokaryocytes was obtained when fusion occurred at an alkaline pH of 8.0 rather than when fusion occurred at pH 7.2. The effect of pH on hybridization is even more dramatic since no hybrid colonies grew at pH 7.2, whereas 60 colonies per 10^6 cells were obtained when the whole process took place at pH 8.0. Rescue of SV40 from transformed hamster fibroblasts through fusion with AGMK cells is also strikingly pH dependent (Table 4) (3). When alkaline medium pH 8.0 or 8.4 is used, the yield of the virus rescued in heterokaryocytes is about 2 logs higher than at pH 7.2 and the number of virus-yielding heterokaryocytes is higher than those at lower pH.

TABLE 3. *The effect of environmental pH of medium on interspecies cell fusion and hybridization in presence of Sendai virus*

	Fusion		Hybridization
pH of medium	% Multinucleated cells	Heterokaryocytes as % of fused cells	Average number[a] of hybrid colonies per 10^6 cells seeded on petri dish
7.2	7.6	31.1	0
8.0	17.2	43.6	60

[a] Recorded 14 days after seeding; fresh medium at the indicated pH added 1, 3, 6, 9, and 12 days after fusion.

TABLE 4. *Effect of environmental pH on rescue of SV40 from transformed hamster cells after fusion with AGMK cells*

pH of medium[a]	Titer of SV40 in cultures 7 days postfusion (infectious units/ml)
7.2	2.5×10^2
8.0	1.1×10^4
8.4	2.5×10^4

[a] At the time of fusion and kept at the same pH after fusion.

This increased virus rescue cannot be explained by increased susceptibility to infection of the permissive AGMK cells at alkaline pH (3).

Turning to other factors which bear somewhat more directly on the mechanism of rescue, it was mentioned before that SV40 coat protein (VP-antigen) can be detected in all the nuclei of heterokaryon cells formed between SV40-infected permissive and nonpermissive cells (81). We then postulated that it is the cytoplasm of the permissive cell that synthesizes this antigen, which is then "soaked up" by permissive and nonpermissive nuclei alike. Recently, it was found (35) that all heterokaryons produced between permissive and nonpermissive cells, regardless of the ratio of nuclei of one origin to nuclei of the other, can be infected with SV40.

The fact that the permissive cell nucleus plays no direct role in the rescue of SV40 has now been amply documented in experiments on fusion of SV40-transformed cells with either enucleated (13), mitomycin-C treated, or irradiated AGMK (Table 5) permissive cells (35). These fusions result in production of infectious virus. Thus, it is the "cytoplasmic broth" in the absence of any nuclear function of the permissive cell that amplifies the replication capability of the already rescuable virus from the nonpermissive

TABLE 5. *Inquiry into the mechanism of virus activation in heterokaryocytes formed between SV40-transformed cells and normal cells*

Treatment of cells prior to fusion	Percent of heterokaryocytes yielding virus on fusion	
	Pretreatment of nonpermissive transformed[a]	Pretreatment of permissive normal[b]
None	0.1–1	
Irradiation	0	0.1–1
Mitomycin C	0	0.03–0.3
Cytochalasin B:		
Enucleation	0	0.1–1
Multinucleation	0.2–3	0.2–2

[a] Mouse cells.
[b] AGMK (African green monkey kidney cells) culture lines.

cell. In contrast, the presence of a biologically functioning nucleus in the transformed nonpermissive cell is essential for rescue of the virus since virus cannot be rescued from enucleated, mitomycin C treated, or irradiated cells. If the transformed cells were treated with Cytochalasin B (Table 5) at a concentration causing formation of multinucleated cells, the ratio of transformed to permissive nuclei after fusion with permissive AGMK cells increased approximately threefold and more heterokaryocytes appeared to yield virus.

One of the hypotheses put forward to explain the mechanism of rescue postulates "random cleavage of the transformed cell DNA" mediated by endogenous cellular nucleases or physical shearing (2). If the results of isolation of infectious SV40 DNA from extracts of transformed cells (2) are confirmed and this hypothesis for the rescue of DNA accepted, one could further postulate that the effect of pH, Cytochalasin B (67), and Sendai virus (85) is to increase the random cleavage of the cell DNA and thus improve the chances for the release of SV40 DNA from integration. Experiments in progress may confirm or repudiate the above hypothesis, but whatever the results, they will enlarge our knowledge of the mechanisms involved in the rescue of viruses from a latent state, a phenomenon in which somatic cell fusion plays such a prominent role.

REPRESSORS REPRESSORS

The metaphor "what is good for *E. coli* is good for the elephant" will have to be modulated if applied to repressors for virus replication in transformed animal cells.

Although a repressor for lambda phage has been identified (72), there is no evidence (37) of the existence of the same type of repressor in SV40-transformed cells (5). There are at least three sets of experimental data that refute the existence of a repressor in SV40-transformed cells (Table 6). In the early studies (37), SV40-transformed, nonpermissive cells were fused either with normal permissive cells or with permissive cells transformed by SV40. Rescue of infectious SV40 occurred in all four types of heterokaryocytes, indicating the absence of a repressor in the permissive transformed cells (Table 7). Further, rescue of SV40 from nonpermissive SV40-transformed cells after their cocultivation with enucleated permissive cells (13) or their fusion with irradiated permissive cells (35) seems not only to repudiate the existence of a repressor for SV40 function but to indicate that a preformed product in permissive cell cytoplasm is all that is necessary to induce SV40 rescue. Finally, some permissive transformed cells were thought resistant to superinfection with complete SV40. Following exposure of these seemingly resistant cells to SV40 DNA, infectious virus was synthesized. The resistance to superinfection presumably was due to a surface change in the SV40-transformed cell whereby the cells were rendered virus resistant (84).

The absence of a repressor in the SV40-transformed cells does not mean that the virus-host association in other tumor viruses may not involve some sort of a repressor. An adenovirus inhibitory factor (for permissive cell infection) was isolated from adenovirus-12-induced hamster tumor cells (8).

In the study of RNA tumor viruses, the restrictive control over replica-

TABLE 6. *Lack of repressor in SV40-transformed cells demonstrated by:*

1. Rescue of SV40 in heterokaryons formed by fusion of SV40-transformed nonpermissive cells with SV40-transformed permissive cells.
2. Rescue of SV40 from nonpermissive transformed cells after contact with enucleated permissive cells or fusion with irradiated permissive cells.
3. Superinfectibility of permissive transformed cells by SV40 or SV40 DNA.

TABLE 7. *Isolation of infectious SV40 from heterokaryocytes formed after fusion of nonpermissive SV40-transformed cells with permissive SV40-transformed cells*

Nonpermissive SV40-transformed cells	Fusion with cells	Isolation of infectious virus
Mouse or Hamster	Normal AGMK	+
	AGMK SV40 transformed	+
	Normal human	+
	Human SV40 transformed	+

tion of murine leukemia virus exercised by the human component of human-mouse heterokaryons may reaffirm the postulate of existence of a "repressor" in human cells which inhibits oncorna virus replication (89).

INDUCED POLYKARYOCYTOSIS AND CHANGED REACTIVITY OF CELLS TO VIRUS INFECTION

Poliomyelitis virus is infectious only for human and some primate cells. Although the uptake of the complete virus by cells of other species is blocked, they can be infected by infectious polio ribonucleic acid. Since the cell surface of the nonpermissive cell is the barrier to viral infection, Sendai-induced fusion of the nonpermissive cells apparently permits their infection through some derangement of the cell membrane (18). In a later study (64) it was shown (Table 8) that polykaryocytes of nonpermissive hamster cells formed after treatment with β-propiolactone-inactivated Sendai virus were not only capable of synthesizing new virus but also showed the cytopathic effect characteristic of permissive cells. Moreover, virus harvested from the polykaryocyte cultures could be passaged serially in the same type of culture if a large enough inoculum was used to infect the cells.

A similar approach was used in the study of human cells nonpermissive to polyoma infection (57) (Table 8). No polyoma replication occurred in these cells when they were fused by Sendai virus prior to the adsorption of the polyoma virus. If, however, 2.5 hr after adsorption of the polyoma virus, the cells were exposed to inactivated Sendai, the polykaryocytes showed marked morphologic alterations and were able to synthesize new virus. In the infected astrocytes the morphological alterations consisted of disappear-

TABLE 8. *Changed reactivity of animal cells to viruses after Sendai virus treatment of the nonpermissive cells*

Virus	Nonpermissive cells		Results: Before exposure to Sendai	Results: After exposure to Sendai
Polio	Hamster embryo line	Replication	One cycle after exposure to viral RNA	Several cycles
		Lesion	None	Cytolysis
Polyoma	Human astrocytes, fibroblasts, Hela	Replication	None	One cycle
		Lesion	None	Cytolysis[a]
Rous Sarcoma	Human embryonic skin muscle cultures	Replication	None?	?
		Lesion	None?	Transformation

[a] Transformation of human embryonic fibroblasts reported in one paper.

ance of the cell processes, pyknosis of the nuclei, and detachment of the cells. These changes occurred first in the polykaryocytes and then in other cells in the culture. No such cytopathic changes occurred in astrocytes exposed only to polyoma virus or Sendai virus alone, or if they were treated with Sendai virus prior to exposure to polyoma virus. No morphological transformation of polyoma-infected astrocytes was observed, although in another report (76) human embryonic fibroblasts pretreated with Sendai virus and then exposed to polyoma were found to transform. The same cultures became transformed when, following absorption of Rous Sarcoma virus of chickens, they were exposed to Sendai virus.

SUSCEPTIBILITY OF HYBRID CELLS TO CHALLENGE WITH VIRUSES

Since hybrids between SV40 permissive and nonpermissive cells are resistant to challenge with SV40 (84) (Table 9), 19 man-mouse and four AGMK-mouse hybrid cell lines were exposed to SV40 DNA in order to bypass possible cell-surface resistance (48). Only three of the man-mouse hybrids supported viral replication, and all three represented hybrid progeny of the same SV40-transformed human cell (W18Va2), which is itself susceptible to superinfection with SV40 DNA. All hybrids, with the exception of one line, had lost a considerable proportion of primate chromosomes by the time of their exposure to SV40 DNA. Because of the selective system used, human X and E-17 chromosomes were known to be present in some of the hybrids; other human chromosomes present either in the resistant or in the susceptible hybrid could not be as easily identified at the time when

TABLE 9. *Challenge of hybrid cells with viruses*

Infecting virus	Parental cells: Permissive	Parental cells: Nonpermissive	Expression of late viral function
SV40-DNA	human	mouse	Only if human chromosome C-7 and 1 or 2 others are present
Polio	human	mouse	Yes for some clones[a] None for some clones
Polyoma	mouse	human	Yes
	mouse	hamster	Yes[b]
Adeno-2 and -12	human	mouse	None after human chromosome loss
Moloney leukemia	human	mouse	None until human chromosome loss

[a] Clones contained more human acrocentric chromosomes.

[b] Decreases as the number of mouse chromosomes proportional to hamster chromosomes decreases.

these experiments were performed. Subsequently it was found that the integration site of SV40 in human cells is on the C-7 chromosome (14). If none of these human chromosome reduced hybrid cells contained the human C-7, it is possible that successful viral replication was unable to occur; hybrids derived from the SV40-transformed parental cell all contained T antigen (hence the C-7 chromosome), and for this reason virus replication could occur.

The lack of techniques for the positive identification of human chromosomes at that time also made it impossible to assign the receptor gene for polio virus to a particular human chromosome (53). No differences were observed in the number of biarmed chromosomes of human origin in polio-resistant and polio-sensitive man-mouse hybrid cells; however, the total number of acrocentric chromosomes was lower by at least four in the resistant cell suggesting the involvement of one of them in coding for receptor activity. The receptor for polio virus is thought to be coded for by a different chromosome than that for nine other picorna viruses tested. Human-mouse hybrid cell lines resistant to polio virus by dint of selective killing of cells bearing the polio receptor were still capable of absorbing the rhino, ECHO, and Coxsackie-B viruses used indicating that genes coding for receptor activity for these viruses and polio are not linked (60).

The yield of virus in man-mouse hybrids challenged with polyoma virus, for which human cells are nonpermissive, was comparable to that obtained after infection of permissive parental mouse cells (71). Hamster-mouse hybrid cells also remained as susceptible to polyoma infection as the parental mouse cells as long as the balance between the mouse chromosomal complement and hamster complement was maintained. Increased amounts of hamster genetic material decreased susceptibility of the hybrid cell to polyoma infection, but only when this was accompanied by a considerable reduction in mouse chromosomes did the cell become completely resistant to polyoma infection (1).

A population of man-mouse hybrid cells, of which 40% was found to synthesize V antigen of SV40 (71), was completely resistant to challenge with either adeno-2 or -12 viruses. Thus, in this case, the hybrid cells, containing at least 23 human chromosomes, behaved like the nonpermissive parental mouse cells. An interesting interaction of adeno-12 virus and man-mouse hybrid cells has recently been reported. This virus induces chromosome gaps in a presumably random manner in the mouse chromosomes in the hybrid cell and in a nonrandom manner in the human chromosomes in these same cells. Previous evidence from adeno infection of human cells *in vitro* had shown that one gap is induced in the proximal part of the long arm of the E-17 chromosome. These workers showed that the gene for thymidine kinase is closely associated with this gap; indeed the amount of this enzyme is increased on infection with adeno-12. Adeno virus-induced deletions on the long arm of E-17 were used to localize the thymidine kinase locus between the distal region of 17q21 and 17q22 (59).

Human-mouse hybrid cells which have lost at least half of their human chromosomes were found to be permissive for synthesis of Moloney leukemia virus (MLV) although human cells and man-mouse heterokaryons were found to be resistant to infection. These data support the hypothesis that a human gene(s) may control a repressive function for MLV synthesis. Loss of a hitherto unidentified chromosome(s) containing this gene may result in the ability of the hybrid cell to replicate the virus (89).

Taking into consideration all experiments on virus-cell interaction in heterokaryons and hybrid cells a small ray of light is beginning to be shed on the genetic aspects of infection with viruses. Integration of the SV40 genome to a particular human chromosome has been found with cosegregation of the T and tumor-specific transplantation antigens (TSTA) with the human C-7 chromosome (14). Viral-induced chromosome lesions are beginning to be exploited. Suppression of viral-induced antigens and virus replication is also beginning to be investigated. Through these types of experiments, the mechanism of transformation by oncogenic viruses and the expression of malignant properties by the transformed cells are being explored.

EXPRESSION OF CELL SURFACE ANTIGENS IN HYBRID CELLS

Cell surface antigen expression was one of the first parameters investigated in hybrid cells. In diploid somatic cells derived from a hybrid animal, the isoantigens inherited from two immunologically distinct parents are fully or codominantly expressed. The original investigations asked whether this was also true of tetraploid, artificially created, somatic cell hybrids. The major murine histocompatibility (H-2) antigens and two non-H-2 antigens were investigated in C3H (H-2^k) and polyoma transformed Swiss mouse hybrid cells (such as H-2^q) and found to be fully expressed (79). Presence of cell surface histocompatibility-type antigens on hybrid cells has since been utilized by a number of investigators for ascertainment of hybrid cells. Later studies showed that interspecific mammalian cell heterokaryons and hybrid cells expressed species-specific antigens derived from each of the parents. A sensitive mixed-immune hemadsorption system was devised for the mouse-human systems. Direct visualization of the glass-attached, tissue-cultured cells coated with an indicator system (antibody-coated red blood cells, antibody-coated bacteria, or both in the double mixed-immune adsorption) showed that antigens of both parental species fairly evenly mixed in heterokaryocytes. The rapidity of the distribution of the surface antigens in heterokaryocytes was remarkable in that 1 hr after fusion of HeLa cells with murine Ehrlich ascites tumor cells (EAT), mouse antigens were found over the entire cell surface. The two parental-species surface antigens continued to be expressed in the mixed HeLa-Ehrlich heterokaryocyte for a prolonged time (93).

Further refinement in the techniques for the demonstration of antigens on mouse-human hybrids and heterokaryons utilized detection of species antigens by murine alloantisera and by rabbit antihuman serum reacting with a mixture of fluorescein-labeled and rhodamine-labeled antiglobulins (23). Fluorescent microscopy utilizing the proper filters then allows classification of cells into green (fluorescein tagged: mouse), red (rhodamine tagged: human) or mosaic (red-green mixture: both). This system was used to determine the mechanism of spread of the surface antigens in heterokaryons immediately postfusion; it was found that, after 25 min of incubation at 37°C following treatment with Sendai virus, 50% of the heterokaryons were total mosaics for the two fluorochromes; by 40 min post-transfer, 90% were total mosaics. The rapidity of the intermixing and the refractory nature of this intermixing to inhibitors of protein synthesis implies that no new protein synthesis is required. Rather there exists a pool of previously synthesized antigen subunits; the fluidity of the cell surface of a heterokaryon is implicated, with either diffusion within the plane of the membrane or the movement of existing antigen from one membrane site into the cytoplasm and reemergence at a new membrane site. This progress is inhibited by temperature reduction and the authors infer this effect is due to a change in the viscosity of the lipid of the cell membrane as low temperatures affect diffusion (23).

Mouse-human hybrid cells were investigated in order to correlate the presence of human chromosomes with expression of human surface antigens (Table 10). Murine antigens are uniformly expressed on these hybrid cells. After human-chromosome loss, from a series of hybrids, a somewhat linear relationship between the number of cells positive for human antigen in a mixed hemagglutination reaction and the number of human chromosomes in the particular clonal population under study was found. If less than five human chromosomes were present, it was still possible to distinguish the clones from the mouse parental line but with less certainty than when more than six human chromosomes were present (97). When hamster-murine hybrid cells were studied by isotopic antiglobulin binding and mixed hemadsorption (74) using antisera against nuclei of the two species antigens, it appeared that the genetic loci specifying immunogenic membrane components were spread throughout the chromosomes of the cells. In the hybrid investigated the best estimate of the proportion of mouse chromosomes was 61%, and 63% of the cell-surface antigens appeared to be of murine origin. Thus no single antigenic component would constitute a large proportion of the membrane surface. Investigations in our laboratories showed that human-rat and human-hamster hybrid cells exhibited both parental species-specific antigens as did the human-murine hybrid cells (Table 10). Some of the clonal populations of human-murine cells did react with rabbit antisera to human cells and some did not. Isoantisera from multiparous women detected antigens on the surface of some of these hybrid clones which were most probably HL-A (the major human histocompatibility antigen) (40). All clones expressed the murine histocompatibility antigens. Others using mouse-human hybrid clonal lines that had lost substantial numbers of human chromosomes raised the possibility of linkage of a human antigen detected by the cytotoxic action of rabbit antihuman serum and complement with one of the genes coding for human lactic dehydrogenase (LdH) (63) (Table 10). Chinese hamster–human hybrid cells also lose human chromosomes (73), and a human autosomal linkage relationship was detected between complement-mediated cytolysis of cells grown in 1% rabbit antihuman cell sera (presence of a human lethal antigen A_L) (Table 10) and the gene controlling the synthesis of the LdH-A subunit. As LdH-A has been assigned to the C-11 chromosome (75), the lethal antigen A_L must also be on this chromosome. Another lethal antigenic marker B_L, which does not cosegregate with A_L, has recently been described (100). The linkage group of this antigen has not yet been determined.

Clonal populations of mouse-human hybrid cells produced through fusion of human peripheral white blood cells derived from HL-A-typed donors with mouse cells (41) (Table 10) were investigated using HL-A-typing sera. When the human donor was HL-A2, 5, and 9, the hybrid clones tested from this fusion contained either no HL-A specificities or only HL-A9, implying the retention of only one of the chromosome pair coding for HL-A. One hybrid clone was found that contained a species-specific antigen detected by heterologous sera to human cells but which was HL-A negative. Thus HL-A and some species-specific antigens may not be linked (41). Preliminary data

TABLE 10. *Expression of human cell-surface antigens in hybrid cells*

Human antigen investigated	Hybrids derived from human X	Results
Species	Rat hamster mouse	Mostly present in mixed populations not detected in some clones of mouse-human hybrid
Lethal antigen (A_L)	Chinese hamster	Linkage with human LdH-A[a,b]
Cytolytic antigen	Mouse	Linkage with human LdH[b]
HL-A	Mouse	Not linked with some species antigens Claims for linkage with P, phosphoglucomutase-3 Tentatively assigned to chromosome C-6

[a] Assigned to C-11.
[b] LdH = Lactic dehydrogenase.

from our laboratory indicate that HL-A and LdH-A and B are not linked (49), suggesting that A_L and the cytotoxic antigen detected by Miggiano et al. are not HL-A antigens. Another group has studied HL-A in human mouse hybrids (19). Utilizing a human diploid fibroblast of the haplotypes HL-A2, W18, and Ao28 Da18, they have isolated four types of clones: those containing antigens of neither haplotype (17 clones), those containing only the Da18 (2 clones), those containing HL-A2 and not Da18 (6 clones), and those containing antigens of both haplotypes (10 clones). Therefore, random segregation of HL-A chromosomes in the hybrid cells occurs. The P antigen system seems to segregate with the HL-A system (19). A linkage group for the HL-A gene has been generated with tentative assignment to the human C-6 chromosome. The linkage group could contain the genes for phosphoglucomutase-3 and HL-A (54, 49).

EXPRESSION OF TUMOR VIRUS ANTIGENS IN HYBRIDS AND HETEROKARYOCYTES

The presence of viral-induced antigens has also been investigated in hybrid cells (Table 11). Tumor induction by hybrid cells produced on fusion of polyoma-transformed mouse cells with L cells was somewhat inhibited when the injected mice were previously immunized with polyoma virus. Polyoma (Py) tumor specific transplantation antigen (TSTA) therefore seemed to be present in these cells (24). Different intraspecific murine hybrid cells (Table 11) were used to preimmunize animals and tumor formation was sup-

TABLE 11. *Expression of tumor virus antigens in hybrids*

Transforming virus	Parental cells	Expression: T-antigen	Expression: Tumor specific transplantation antigen	Expression: Viral protein antigen
Polyoma	Mouse[a] × mouse tumor	+	+	
	Mouse[a] × mouse fibroblasts	+	+	
	Mouse[a] × L cells		+	
Simian virus 40	Human[a] × mouse	[b]+ clones and − clones	+[b]	+[b]
			−	−
	Rat[a] × mouse	+		
Epstein-Barr	Human[a] × human			−
	Human[a] × human (IUDR)			+
	Human[a] × mouse	[c]+ clones and − clones		

[a] Transformed cells.
[b] All + clones contain human chromosome C-7.
[c] EBNA – Epstein-Barr nuclear antigen.
+ = Presence of antigen
− = Absence of antigen

pressed when these mice were later challenged with an isologous Py-induced tumor (TSTA was present). Py-induced, complement-fixing nuclear antigen (T antigen) was also found in these hybrid cells (15). TSTA and T antigens were both found (16) in hybrids (Table 11) formed between polyoma-transformed mouse cells and normal diploid mouse cells. Indirect immunofluorescence (Table 12) was used to visualize SV40-induced T antigen on the nuclei of heterokaryons formed either by Sendai virus–induced fusion of SV40-transformed cells with nontransformed cells (80) or by incorporation of nuclei from cocultivation of SV40-transformed human cells with mononucleated rat myoblasts at the time of development *(in vitro)* of multinucleated myotubes (21). All nuclei present in the heterokaryocyte were found to contain SV40-induced T antigen within 24 hr of fusion. This T antigen transfer was found to be dependent on RNA and protein synthesis in the heterokaryocyte with uptake of the newly synthesized protein by all nuclei (whether permissive or nonpermissive for SV40 synthesis) within the heterokaryocyte. The SV40 coat protein (VP) was also demonstrated by immunofluorescence on all of the nuclei in heterokaryocytes made between preinfected SV40 permissive cells and nonpermissive cells. Apparently, both SV40-induced T antigen and the VP are synthesized in the common cytoplasm and then taken up by any nuclei in the heterokaryon (81). A similar result was found in human-hamster heterokaryons infected with either adeno-12 or adeno-2 viruses (95). These viruses grow only in human cells and cannot replicate or induce inclusion bodies (IB) in hamster cells. Hamster nuclei, however, in heterokaryons showed a high incidence of viral IB. The appearance of the adenovirus-induced IB in the nucleus is once again a nonspecific process (95).

TABLE 12. *Expression of tumor virus antigens in heterokaryons*

		Expression	
Virus	Parental cells	T antigen	Viral protein antigen
SV40	AGMK[a] × mouse	+	+
	AGMK[a] × hamster	+	+
	AGMK[a] × AGMK[a]	+	+
Adeno-2[b]	Human × hamster		+
Adeno-12[b]	Human × hamster		+
Mammary tumor	Mouse[a] × mouse		++++
	Mouse[a] × rat		+
Moloney leukemia	Mouse[a] × human		−

AGMK = African green monkey kidney.
[a] Cells infected with or transformed by the virus prior to fusion.
[b] Heterokaryons challenged with virus.

When cells shedding mouse mammary tumor virus (MTV) (Table 12) were fused with mouse cells and the mixed population of hybrids, heterokaryons, and parental cells were investigated by immunofluorescence for the presence of the virus, a higher proportion of cells in the fused population (63 to 84%) was positive than that found in the parental population (10 to 25%). This observation was confirmed by electron microscopy as 20 to 30% of the parental cells contained B particles in comparison to 68 to 85% of the fused population. Fused populations of the MTV cells with rat mammary gland-derived cells did not exhibit this increased MTV yield and a species-specific effect was seen as only the mouse cells in the mixed population were involved in viral replication (55).

Hybrid cells formed between SV40-transformed human cells (T antigen positive) and nonpermissive murine cells were investigated to determine the presence of T antigen on hybrid nuclei (Table 12). It was initially found that the presence of human chromosomes was necessary for the presence of T antigen on the hybrid cell nucleus (98, 99). Utilizing interspecific human-mouse hybrid cells, a linkage relationship between the human C-7 chromosome and the SV40 genome has recently been established. Concordance of the C-7 chromosome and SV40 T antigen, TSTA, and viral protein synthesis after rescue has been seen not only in interspecific but also in intraspecific human hybrid cells (14).

All clonal lines of hybrid cells derived from fusions of SV40-transformed rat cells and normal mouse cells showed SV40 T antigen. Chromosome loss was minimal in this combination (65).

Oncorna virus–induced antigens appear in the majority of mouse cells after infection with MLV while MLV-exposed human fibroblasts never express this antigen. Heterokaryons between mouse and human fibroblasts just after infection of the mouse cells with MLV never express the viral antigen, although the mouse synkaryons in the population behave positively (Table 12).

Intraspecific hybrid cell lines derived from fusions between a virus-shedding Burkitt lymphoma–derived lymphoblastoid cell (EB) and a human sternal marrow cell line (Table 12) showed no evidence of expression of the viral genome by conventional immunofluorescent techniques, although they contain virus-specific DNA as determined by DNA-RNA hybridization (26). These hybrids can be induced to express the EBV genome (i.e., EBV-specific antigens and visualization of virus by electron microscopy), by treatment of the cells with halogenated pyrimidines (25). These data suggest that the herpes virus genome is integrated in these nonlymphoblastoid cell lines and this integration is reversible by drug treatment. Attempts to correlate presence of EBNA (Epstein-Barr nuclear antigen) with a particular human chromosome(s) are now in progress in order to find the EBV integration site.

ANTIGEN EXPRESSION AND MALIGNANCY IN INTRASPECIFIC MURINE HYBRIDS

A series of papers have been published correlating the presence or absence of antigenic expressions in various hybrid cell lines derived from tumorigenic parental cells (Table 13). The Ehrlich ascites tumor (EAT) plays a unique role as a parental cell in fusion reactions since not only is there no known H-2 antigen expressed in these cells (9, 94), but it also causes the suppression of the L cell antigens [H-2^k, Friend Moloney Rauscher type antigen (FMR) and the antigen of the C type particles associated with L cells (L)] in EAT-L cell hybrids (30, 33, 44). The EAT cell is also the most efficient parental cell at suppressing the Fc-specific receptor in macrophage cells (see below) (28). That antigen suppression is unique to EAT was shown by comparing hybrids between L cells and EAT with those made by fusing L cells and an ascites sarcoma induced by polyoma (SEWA) and L cells and an ascites sarcoma induced by methylcholanthrene (MSWBS). Contrary to the results with L-EAT, in both the L-SEWA hybrids and the L-MSWBS hybrids there was expression of the H-2^k and FMR antigens of the L cell parent antigens (33). When clonal populations of EAT-L hybrid cells were investigated, a segregant was found that fully expressed the H-2^k phenotype; therefore the gene coding for H-2^k was present in the initial hybrid population and its expression was suppressed. Malignancy and H-2 expression are apparently unrelated as most L-EAT hybrids have a low level of malignancy as do L-SEWA and L-MSWBS hybrids; some L-EAT hybrid clones with a high degree of malignancy strongly express the L cell H-2^k locus (44).

Further study of L-EAT hybrid clones from an original hybrid population with a low level of malignancy and suppression of the expression of the three L cell derived antigens shows reappearance of the three cell-surface antigens occurring independently of each other and independently of the malignancy of the hybrid clone. Loss of a number of chromosomes is linked with the expression of malignancy and with the presence of each of these antigens (30).

Investigation of the relationship between cell-surface antigens and the release of Moloney type leukemia virus from hybrid cells utilized two different Moloney-induced murine lymphoma cell lines (Table 13). One line (YAC) releases large quantities of virus and expresses the Moloney-specific surface antigen, and one (YACIR) almost completely resists anti-Moloney serum cytotoxicity and does not release infectious Moloney virus. L cell hybrids with each of these lymphoma lines expressed Moloney-specific surface antigen to the same degree, contained both parental H-2 isoantigens on their surfaces, and possessed the L antigen. Infectious virus was released only by the L cell-YAC hybrid. The YACIR parental line does not behave like EAT in suppression of Moloney virus-induced surface antigens in

TABLE 13. *Expression of antigens in hybrid cells*

Parental cells*	Results			
	Histocompatibility antigens	Other antigens	Malignancy	Virus release
EAT × L	Suppressed in uncloned	Suppression of "L"** and FMR**	Low	ND
	In some clones $H\text{-}2^k$ present	"L" and FMR segregate independent of each other and of H-2	High	ND
SEWA × L	$H\text{-}2^s$ and $H\text{-}2^k$	"L"	Low	ND
MSWBS × L	$H\text{-}2^s$ and $H\text{-}2^k$	"L"	Low	ND
YAC × L	$H\text{-}2^d$ and $H\text{-}2^k$	"L" and FMR	Very low	Yes
YACIR† × L	$H\text{-}2^d$ and $H\text{-}2^k$	"L" and FMR	Low	No
TA3†† × normal fibroblasts	$H\text{-}2^a$ and $H\text{-}2^f$	ND	One clone-none Others-high	ND

* EAT = Ehrlich ascites tumor; SEWA-Polyoma-induced ascites sarcoma; MSWBS = Methylcholanthrene-induced ascites sarcoma; YAC = Moloney-induced lymphoma; YACIR = Moloney-induced lymphoma; TA3 = mammary carcinoma.

** L cells contain antigen "L," a C type particle and FMR (Friend Moloney Rauscher leukemia antigen).

† Suppressed: Cell line does not release infectious Moloney virus.

†† Low-expressed: Reduced expression of H-2 antigen.

ND = Not done.

hybrid cells; however, infectious virus particles are not released (20). Expression of the virus-induced cell-surface antigen and release of virus are separately controlled.

Another immunoresistant tumor cell line TA3, which exhibits (Table 13) reduced expression of its H-2 antigen and therefore has a large range of host strains, was fused with mouse diploid fibroblasts and the hybrids were found to contain high levels of both H-2 antigens. These data lend further weight to the unique nature of the EAT cell (45). If one is to take this murine example and apply it to other systems, the original concept of antigen suppression in tumor cells no longer is a generality. One point worth noting is the apparent lack of linkage between malignancy and the antigens associated with a C-type tumorigenic virus. In general, study of the malignancy of hybrid cells derived from fusions of tumorigenic and nontumorigenic cells is open to further investigation. Another relevant experiment showed that intraspecific human hybrid cells which contain close to a tetraploid number of human chromosomes, derived from fusion between a contact inhibited nontumorigenic line and an SV40-transformed (WI38) human cell, retain their transformed phenotype (C. Croce and H. Koprowski, *in press*).

IMMUNE FUNCTIONS IN HETEROKARYONS AND HYBRIDS

Basically, the approach to the study of the immune response has been to fuse cells that are capable of performing an "immune" function with those deficient in this function to see if the resultant heterokaryon or hybrid cells synthesize the molecule in question. The function is then found to be either expressed (under positive control) or nonexpressed (under negative control) and an attempt is then made to correlate this information with the presence of chromosomes representing each of the parental lines. Parameters of the immune response which have been so studied are the synthesis of and response to interferon, the production of various components of complement, immunoglobulin synthesis, and the presence of the Fc-specific receptor of the macrophage.

BIOSYNTHESIS OF INTERFERON

Interferon synthesis was first investigated in chick red blood cell–human diploid fibroblast heterokaryons (31). It was found that the heterokaryons produce chicken interferon (Table 14), although no evidence of this function is found in the normal nucleated chicken red blood cells. An investigation of the sensitivity of these heterokaryons to both chicken and human interferons 4 days after fusion revealed that they were also sensitive to chicken interferon. When compared with the human parental fibroblasts, the heterokaryons were somewhat less sensitive to human interferon, and there was a synergistic effect when both interferons were added (32).

TABLE 14. *Biosynthesis of interferon and susceptibility to its induction in heterokaryocytes and hybrid cells*

Parental cells	Results[a]	
	Heterokaryocytes	Hybrids
Human (+) X chicken (−)	Synthesis of chicken interferon Susceptibility to chicken interferon	ND
Hamster (±) X mouse (+)	ND	Increased synthesis of hamster interferon and susceptible to hamster interferon
Monkey (+) X mouse (+)	ND	Some clones produce monkey interferon but are not susceptible. Some do not produce but are susceptible.
Mouse (+) X man (+)	ND	Human interferon induced antiviral protein assigned to G-21 chromosome.

(+) Denotes biosynthesis and inducibility.
[a] Note: In addition to the type of interferon being synthesized as indicated in the columns all hybrids retain the capability to synthesize and respond to interferon of the parental cell.

Interferon synthesis was also investigated in a hybrid clone, obtained by the fusion of mouse cells with Syrian hamster cells. The mouse cells not only produce interferon in quantity but are also sensitive to its action; the Syrian hamster cells produce little interferon and are relatively insensitive to its action. The hybrid clone was found to synthesize both mouse-specific and hamster-specific interferons. There was a 10-fold increase in the production of hamster interferon when compared with the hamster parent. Additionally, the hybrid cells were sensitive to both mouse and hamster interferon, and the hybrid cell was more sensitive than its hamster parent. These data imply that not only were both genomes fully expressed in the hybrid cell but that the mouse genome promoted the production of interferon by the hamster genome as well as allowing increased sensitivity to its action (4).

Several clones from mouse-monkey hybrid cells were studied for their sensitivity to, and their ability to produce, simian interferon (Table 14). All of the hybrid cells studied made and reacted to murine interferon. The original hybrid population was capable of both producing and responding to monkey interferon. However, the loss of monkey chromosomes from mouse-monkey hybrid cells allowed the selection of two different clonal populations: one that produced simian interferon but did not respond to its action and one that was sensitive to simian interferon but did not produce it. Chromosomal analysis suggests that the gene for the elaboration of monkey interferon is associated with a small subtelocentric chromosome of monkey origin. The gene for interferon sensitivity segregates independently from the production gene and is apparently on a different monkey chromosome (6, 7, 83).

The antiviral protein induced by interferon (hence an indication of interferon sensitivity) has been studied in murine-human hybrid cells. Utilizing the quinacrine mustard banding techniques for chromosome identification, it has been found that the genetic information coding for this protein was located on the human G-21 chromosome (87), and that this species-specific feature is retained in hybrid cells as long as the chromosome coding for these functions is there.

COMPLEMENT SYNTHESIS

Two studies have been made on biosynthesis (Table 15) of complement after fusing cells from animals deficient in one of the components of the complement system with C_5 cells that are not deficient. In the first experiments, spleen cells from a strain of mice deficient in the mouse analogue of C_5 were fused both with kidney cells from a coisogenic nondeficient mouse strain and with nucleated chicken red blood cells. Injections of both mixtures into deficient-strain mice resulted in a temporary restoration of the hemolytic capacity of their sera, whereas the sera of the control mice injected with the parental cells separately were not capable of hemolysis (56). Hybrids be-

TABLE 15. *Biosynthesis of complement and immunoglobulin (IgG) and expression of macrophage functions in heterokaryocytes and hybrids*

Cell product investigated	Parental cells	Results
Complement (C_5)	Mouse (−) X mouse (+) X chicken (+)	Restoration of complement activity in C_5 deficient mouse
(C_4)	Guinea pig (−) X Hela (−)	Synthesis of human C_4 in hybrids
IgG	Myeloma (+) X L cells (−)	Slight anti-DNP activity
	Myeloma (+) X mouse fibroblast (−)	No IgG in hybrid detected
	Myeloma (+)[a] X lymphoma (−)	Some hybrids produce both IgG and free kappa chains and others only free kappa chains
Macrophage receptors	Macrophage (+) X melanocytes (−)	Macrophage specific receptors suppressed in heterokaryocyte unless cells are trypsinized
	Macrophage (+) X EAT	
	Macrophage (+) X L cells (−)	Macrophage specific receptors ATPase activity suppressed in hybrids

(+) Denotes biosynthesis of product under investigation.
[a] Produces free kappa chains.
EAT = Ehrlich ascites tumor.

tween peritoneal exudate cells from a guinea pig strain deficient in the ability to synthesize the fourth component of complement (C_4) and HeLa cells that do not synthesize C_4, synthesize C_4 immunologically identified as human. The guinea pig chromosome complement was much reduced. It appears that either a guinea pig chromosome that repressed C_4 production in the guinea pig cell was lost as was the guinea pig chromosome containing the C_4 structural gene or the ability to produce C_4 may have been initiated in hybrid cells by the guinea pig genetic information. In the latter case, the defect in the guinea pig would be a mutation of the structural gene (11). More information on this point is provided in this volume.

IMMUNOGLOBULIN SYNTHESIS

The regulation of immunoglobulin synthesis in hybrid cells (Table 15) produced by fusing an immunoglobulin-synthesizing mouse myeloma cell with a mouse cell that does not express this differentiated function has been investigated in three hybrid populations (10, 61, 62, 69). One group (69) utilized a hybrid between an L cell and a myeloma that secretes an anti-DNP immunoglobulin with A-type heavy chain and lambda-type light chain. A small percentage (approximately 6%) of the DNP-binding activity of the plasmacytoma secretion was detected in the supernatants of one hybrid population tested. No immunoglobulin was detectable by Ouchterlony techniques, but radio immunoelectrophoresis identified A-type heavy chain, lambda-type light chains, and anti-DNP activity in the secretion. It was postulated that these glass-attached hybrid cells might be unable to secrete large amounts of immunoglobulin due to their altered membrane structure *in vitro*.

A second report (10) utilized six colonies of a hybrid (Table 15) between a myeloma cell capable of producing both heavy and light chains and a Swiss mouse fibroblast incapable of such production. No immunoglobulin was secreted into the tissue culture medium or within the cytoplasm of these hybrid cells. Direct immunofluorescence using FITC tagged antimouse gamma globulin also showed all cells to be negative. The immune precipitation and electrophoretic techniques used were shown to be capable of detecting heavy and light chains if only 1% of the amount produced by the myeloma parent were present. Loss of the chromosomes containing the genes for both heavy and light chains seemed unlikely as all of the hybrid colonies were tested soon after fusion and, even at the end of the study, the hybrids retained over 80% of the chromosomes theoretically expected if each of the parents contributed their modal chromosome number to the hybrid cell. A third investigation (61, 62) using hybrid cells derived from another Balb/c myeloma (secreting both immunoglobulin and free kappa chain) parental cell fused with a C57B1 lymphoma cell demonstrated synthesis in some hybrid clones of both the parental products and in others of only the free kappa chain (Table 15). Chromosome numbers are higher in those cells secreting both intact immunoglobulin G and free kappa chains. We suggest that immunoglobulin synthesis in this system is neither an inducible nor repressible system but is controlled simply by the presence of chromosomes containing the structural gene for the synthesis of heavy and light chains. If the system were completely inducible, then immunoglobulin synthesis of the nonproducing parent should be turned on. If it were repressible, then immunoglobulin synthesis of the myeloma parent should have been shut off by the nonproducing parent. An alternative exists if one assumes that the system is inducible-repressible but allotype-specific, or that there is a mutation in the operator region of the myeloma parent that results in inability to bind repressor. Either of these explanations may explain the divergent results of the three groups.

More recently lambda-chain synthesis has been found in two clones of a murine-human hybrid cell derived from fusion of a human lymphoblast cell line that only produced the lambda chain. These data suggest that the structural genes for immunoglobulins can be mapped in hybrid cells (68).

FUNCTIONS OF MACROPHAGES

Heterokaryons formed between macrophages and melanocytes were investigated for the ability to perform several differentiated functions of the macrophage (27) (Table 15). Four macrophage-specific functions were studied: the specific plasma-membrane receptors that confer the ability to ingest antibody-coated particles on macrophages, the highly active divalent cation-dependent ATPase, the high amount of the lysosomal acid phosphatase, and the high proportion of refractile lipid droplets. These markers could be discerned in heterokaryons shortly after fusion but gradually disappeared over the 4 to 5 day course of observation, never to reappear. Trypsinization of the heterokaryon cell surface after the macrophage-specific receptors have disappeared results in their reappearance. Inhibition of protein and RNA synthesis and UV irradiation of the nonmacrophage parental cell before fusion eliminates the masking of the macrophage receptors. This suggests that masking results from active synthesis by the nonmacrophage genome in the heterokaryon. Of a number of cell types tested, the Ehrlich ascites tumor cells and melanoma cells were most effective in this masking ability. The authors suggest that masking of cell-surface antigens by tumor cells may allow for tumor growth due to the escape from normal immune surveillance by the host (28). Hybrid cells prepared between mouse peritoneal macrophages and L cells showed no Fc-specific receptor activity and it was impossible to unmask this receptor by mild proteolysis. Another macrophage-specific function, receptor activity to complement, was also abolished in the hybrid cell. The high ATPase activity of the macrophage parent (100%) was intermediately expressed in the clonally derived hybrid cell population (10 to 23%) compared with the lack of expression in the L cell parent (0 to 2%). The H-2 antigens of both parents were expressed. Since the chromosome number was 85 to 100% of the sum of the two parent cells' chromosome number, we felt that the genes for the receptors are present but not expressed (29).

Thus, of the immunoreactive functions studied in heterokaryons and hybrid cells we can say that interferon is always expressed; complement production appears to be a dominant feature if one of the parents is already synthesizing it; the regulation of immunoglobulin synthesis is under some anomalous form of control, which needs much more effective study before drawing any conclusions; and the macrophage-specific receptors, which confer upon this cell its specialized function in the immune system, are not expressed in a dominant fashion. Investigations utilizing systems in which there is preferential loss of one species' chromosome such as mouse-human hybrids that lose human chromosomes may eventually lead to the mapping of the genes controlling all of these functions. Detailed investigations of the mechanisms of the control of immunoglobulin synthesis will be the most interesting avenue of approach in this general area. The key to the macrophage-specific functions awaits further knowledge about the nature of the cell membrane; perhaps the clue to the state of flux of the cell membrane is in the study of membranes of somatic cell hybrids.

At this point, you may be wishing for a less complicated story than the one you have just heard. A wealthy Irish lady who was a friend of James Joyce had certain difficulties in understanding *Ulysses*. She asked Joyce for help and he explained to her what Molly Bloom, Leopold Bloom, and Stephen Dedalus were up to that day. "I said to him," she recalled, " 'Jimmy, it's a lovely story. Why didn't you write it just the way you told it to me?' " A simple story was not the novel Joyce wrote; the topic we are investigating today does not lend itself to easy answers.

ACKNOWLEDGMENTS

This investigation was supported in part by U.S. Public Health Service Grants CA 04534 and CA 10815 from the National Cancer Institute, RR 05540 from the Division of Research Resources of the National Institutes of Health, and funds from the National Foundation—March of Dimes.

REFERENCES

1. Basilico, C., Matsuya, Y., and Green, H. (1970): The interaction of polyoma virus with mouse-hamster somatic hybrid cells. *Virology 41*:295–305.
2. Boyd, V. L., and Butel, J. S. (1972): Demonstration of infectious deoxyribonucleic acid in transformed cells. I. Recovery of Simian virus 40 from yielder and non yielder transformed cells. *J. Virol. 10*:399–409.
3. Calothy, G., Croce, C. M., Defendi, V., Koprowski, H., and Eagle, H. (1973): Effect of environmental pH on rescue of Simian virus 40. *Proc. Nat. Acad. Sci. 70*:366–368.

4. Carver, D. H., Seto, D. S., and Migeon, B. R. (1968): Interferon production and action in mouse, hamster and somatic hybrid mouse-hamster cells. *Science 160*:558–559.
5. Cassingena, R., and Tournier, P. (1968): Mise en évidence d'un "répresseur" spécifique dans des cellules d'espèce différentes transformées par le virus SV40. *C.R. Acad. Sci. Paris 267*:2251–2254.
6. Cassingena, R., Chany, C., Vignal, M., Estrade, S., and Suarez, H.-G. (1970): Utilisation des cellules hybrides (Singe-Souris) pour l'étude de la régulation cellulaire de la production et de l'action de l'interférone. *C. R. Acad. Sci. Paris (D) 270*:1189–1191.
7. Cassingena, R., Chany, C., Vignal, M., Suarez, H., Estrade, S., and Lazer, P. (1971): Use of monkey-mouse hybrid cells for the study of the cellular regulation of interferon production and action. *Proc. Nat. Acad. Sci. 68*:580–584.
8. Champe, P. C., Strohl, W. A., and Schlesinger, R. W. (1972): Demonstration of an adenovirus-inhibitory factor in adenovirus-induced hamster tumor cells. *Virology 50*:482–494.
9. Chen, L., and Watkins, J. F. (1970): Evidence against the presence of H2 histocompatibility antigens in Ehrlich ascites tumour cells. *Nature 225*:734–735.
10. Coffino, P., Knowles, B., Nathenson, S. G., and Scharff, M. D. (1971): Suppression of immunoglobulin synthesis by cellular hybridization. *Nature New Biol. 231*:87–90.
11. Colten, H. R., and Parkman, R. (1972): Biosynthesis of C4 (fourth component of complement) by hybrids of C4-deficient guinea pig cells and Hela cells. *Science 176*:1029–1031.
12. Croce, C. M., Koprowski, H., and Eagle, H. (1972): Effect of environmental pH on the efficiency of cellular hybridization. *Proc. Nat. Acad. Sci. 69*:1953–1956.
13. Croce, C. M., and Koprowski, H. (1973): Enucleation of cells made simple and rescue of SV40 by enucleated cells made even simpler. *Virology 51*:227–229.
14. Croce, C. M., Girardi, A. J., and Koprowski, H. (1973): Assignment of the SV40 T antigen gene to human chromosome C-7. *Proc. Nat. Acad. Sci. 70*:3617–3620.
15. Defendi, V., Ephrussi, B., and Koprowski, H. (1964): Expression of polyoma-induced cellular antigen(s) in hybrid cells. *Nature 203*:495–496.
16. Defendi, V., Ephrussi, B., Koprowski, H., and Yoshida, M. C. (1967): Properties of hybrids between polyoma-transformed and normal mouse cells. *Proc. Nat. Acad. Sci. 57*:299–305.
17. Cited in Eckhart, W. (1972): Oncogenic viruses. *Ann. Rev. Biochem. 41*:503–516.
18. Enders, J. F., Holloway, A., and Grogan, E. A. (1967): Replication of poliovirus I in chick embryo and hamster cells exposed to Sendai virus. *Proc. Nat. Acad. Sci. 57*:637–644.
19. Fellous, M., Billardon, C., Dausset, J., and Frezal, J. (1971): Linkage probable entre les locus "HLA" and "P." *C.R. Acad. Sci. Paris (D) 272*:3356–3359.
20. Fenyö, E. M., Grundner, G., Klein, G., Klein, E., and Harris, H. (1971): Surface antigens and release of virus in hybrid cells produced by the fusion of A9 fibroblasts with Moloney lymphoma cells. *Exp. Cell Res. 68*:323–331.
21. Fogel, M., and Defendi, V. (1968): Internuclear transfer of SV40 T-antigen in absence of infectious virus. *Virology 34*:370–373.
22. Fogel, M., and Sachs, L. (1969): Activation of virus synthesis in polyoma-transformed cells. *Virology 37*:327–334.
23. Frye, L. D., and Edidin, M. (1970): The rapid intermixing of cell surface antigens after formation of mouse-human heterokaryons. *J. Cell Sci. 7*:319–335.
24. Gershen, D., and Sachs, L. (1963): Properties of a somatic hybrid between mouse cells with different genotypes. *Nature 198*:912–913.
25. Glaser, R., and Rapp, F. (1972): Rescue of Epstein-Barr virus from somatic cell hybrids of Burkitt lymphoblastoid cells. *J. Virol. 10*:288–296.
26. Glaser, R., and Nonoyama, M. (1973): Epstein-Barr virus: Detection of genome in somatic cell hybrids of Burkitt lymphoblastoid cells. *Science 179*:492–493.
27. Gordon, S., and Cohn, Z. (1970): Macrophage-melanocyte heterokaryons. I. Preparation and properties. *J. Exp. Med. 131*:981–1003.
28. Gordon, S., and Cohn, Z. (1971): Macrophage-melanoma cell heterokaryons. IV. Unmasking the macrophage-specific membrane receptor. *J. Exp. Med. 134*:947–962.
29. Gordon, S., Ripps, C. S., and Cohn, Z. (1971): The preparation and properties of macrophage-L cell hybrids. *J. Exp. Med. 134*:1187–1200.
30. Grundner, G., Fenyö, E. M., Klein, G., Klein, E., Bregula, U., and Harris, H. (1971): Sur-

face antigen expression in malignant sublines derived from hybrid cells of low malignancy. *Exp. Cell Res. 68*:315–322.

31. Guggenheim, M. A., Friedman, R. M., and Rabson, A. S. (1968): Interferon: Production by chick erythrocytes activated by cell fusion. *Science 159*:542–543.
32. Guggenheim, M. A., Friedman, R. M., and Rabson, A. S. (1969): Interferon action in heterokaryons. *Proc. Soc. Exp. Biol. Med. 130*:1242–1245.
33. Harris, H., Miller, O. J., Klein, G., Worst, P., and Tachibana, T. (1969): Suppression of malignancy by cell fusion. *Nature 223*:363–368.
34. Huebner, K., Croce, C. M., and Koprowski, H. Isolation of defective viruses from SV40-transformed human and hamster cells; unpublished.
35. Huebner, K., and Koprowski, H. *(in press)*: Synthesis of SV40 capsid antigen(s) in heterokaryocytes formed by fusion of nonpermissive with permissive cells. *Virology.*
36. Huebner, K. *(manuscript in preparation).*
37. Jensen, F. C., and Koprowski, H. (1969): Absence of repressor in SV40 transformed cells. *Virology 37*:687–690.
38. Jensen, F. C., Lief, F. S., and Hummeler, K. (1970): Correlation between SV40 rescue and presence of a paramyxo-like virus in SV40-transformed hamster cells. *Fed. Proc. 29*:742A.
39. Jensen, F. C. *(personal communication).*
40. Kano, K., Baranska, W., Knowles, B. B., Koprowski, H., and Milgrom, F. (1969): Surface antigens on interspecies hybrid cells. *J. Immunol. 103*:1050–1060.
41. Kano, K., Knowles, B. B., Koprowski, H., and Milgrom, F. (1972): HL-A antigens on man-mouse hybrid cells. *Eur. J. Immunol. 2*:198–202.
42. Katz, M., Rorke, L. B., Masland, W. S., Barbanti-Brodano, G., and Koprowski, H. (1970): Subacute sclerosing panencephalitis: Isolation of a virus encephalitogenic for ferrets. *J. Infec. Dis. 121*:188–195.
43. Kit, S., Kurimura, T., Salvi, M. L., and Dubbs, D. R. (1968): Activation of infectious SV40 DNA synthesis in transformed cells. *Proc. Nat. Acad. Sci. 60*:1239–1246.
44. Klein, G., Gars, U., and Harris, H. (1970): Isoantigen expression in hybrid mouse cells. *Exp. Cell Res. 62*:149–160.
45. Klein, G., Friberg, S., Jr., and Harris, H. (1972): Two kinds of antigen suppression in tumor cells revealed by cell fusion. *J. Exp. Med. 135*:838–839.
46. Knowles, B., Steplewski, Z., Swetly, P., Barbanti-Brodano, G., and Koprowski, H. (1969): Cell hybridization and tumor viruses: In: *Heterospecific Genome Interaction,* The Wistar Institute Symposium Monograph No. 9, p. 37–49 (ed. V. Defendi), Wistar Institute Press, Philadelphia, Pa.
47. Knowles, B., Jensen, F. C., Steplewski, Z., and Koprowski, H. (1968): Rescue of infectious SV40 after fusion between different SV40-transformed cells. *Proc. Nat. Acad. Sci. 61*:42–45.
48. Knowles, B., Barbanti-Brodano, G., and Koprowski, H. (1971): Susceptibility of primate-mouse hybrid cells to SV40. *J. Cell. Phys. 78*:1–8.
49. Knowles, B., and Koprowski, H. *(unpublished data).*
50. Koprowski, H., Barbanti-Brodano, G., and Katz, M. (1970): Interaction between papova-like virus and paramyxovirus in human brain cells: A hypothesis. *Nature 225*:1045–1047.
51. Koprowski, H. (1971): Cell fusion and virus rescue. *Fed. Proc. 30*:914–920.
52. Koprowski, H. (1972): Considerations of the mechanisms of latency in chronic viral infections of the nervous system. In: *The Neurosciences,* Third Study Program (ed. F. O. Schmitt and F. G. Worden), M.I.T. Press, Cambridge, Mass.
53. Kusano, T., Wang, R., Pollack, R., and Green, H. (1970): Human-mouse hybrid cell lines and susceptibility to polio virus. *J. Virol. 5*:682–685.
54. Lamm, L. U., Sverjgaard, A., and Kissmeyer-Nielsen, F. (1971): PGM_3:HL-A is another linkage in man. *Nature 231*:109–110.
55. Lasfargues, E. Y., Kramarsky, B., and Moore, D. H. (1971): Effects of cell fusion on production of the mouse mammary tumor virus-immunofluorescence study. *Proc. Soc. Exp. Biol. Med. 136*:777–781.
56. Levy, N. L., and Ladda, R. L. (1971): Restoration of haemolytic complement activity in C_5-deficient mice by gene complementation in hybrid cells. *Nature New Biol. 229*:51–52.
57. Levy, N. L. (1972): Synergism between polyoma and inactivated Sendai virus in human astrocytes, fibroblasts and Hela cells. *J. Neuropath. Exp. Neurol. 31*:639–644.

58. Long, C., Kelloff, G., and Gilden, R. V. (1972): Variations in sarcoma and leukemia virus activity in somatic cell hybrids. *Int. J. Cancer 10*:310–319.
59. McDougall, J. K., Kucherlapati, R., Ruddle, F. H. (1973): Localization and reduction of the human thymidine kinase gene by adenovirus 12. *Nature New Biol. 245*:172–175.
60. Medrano, L., and Green, H. (1973): Picornavirus receptors and picornavirus multiplication in human-mouse hybrid cell lines. *Virology 54*:515–524.
61. Mohit, B. (1971): Immunoglobulin G and free kappa chain synthesis in different clones of a hybrid cell line. *Proc. Nat. Acad. Sci. 68*:3045–3048.
62. Mohit, B., and Fan, K. (1971): Hybrid cell line from a cloned immunoglobulin producing mouse myeloma and a nonproducing mouse lymphoma. *Science 171*:75–77.
63. Nabholz, M., Miggiano, V., and Bodmer, W. (1969): Genetic analysis with human-mouse somatic cell hybrids. *Nature 223*:358–363.
64. Neff, J. M., and Enders, J. F. (1968): Poliovirus replication and cytopathogenecity in monolayer hamster cell cultures fused with Beta propiolactone-inactivated Sendai virus. *Proc. Soc. Exp. Biol. Med. 127*:260–267.
65. Noordaa, J. van der., Haagen, A. van, Walboomers, J. M. M., and Someren, H. van (1972): Properties of somatic cell hybrids between mouse cells and Simian virus 40-transformed rat cells. *J. Virol. 10*:67–72.
66. Nyormoi, O., Klein, G., Adams, A., and Dombos, L. (1973): Sensitivity to EBV superinfection and IUdR inducibility of hybrid cells formed between a sensitive and a relatively resistant Burkitt lymphoma cell line. *Int. J. Cancer 12*:396–408.
67. O'Neill, F. J. (1972): Chromosome pulverization in cultured normal and neoplastic cells treated with cytochalasin B. *J. Nat. Cancer Inst. 43*:1733–1737.
68. Orkin, S. H., Buchanan, P. D., Yount, W. J., Reimer, H., and Littlefield, J. W. (1973): Lambda-chain production in human lymphoblast–mouse fibroblast hybrids. *Proc. Nat. Acad. Sci. 70*:2401–2405.
69. Periman, P. (1970): IgG synthesis in hybrid cells from an antibody-producing mouse myeloma and an L cell substrain. *Nature 228*:1086–1087.
70. Periman, P., Tyrrel, S., and Rabson, A. S. (1972): Herpesvirus saimiri in marmoset-mouse hybrid cell lines. *J. Nat. Cancer Inst. 49*:387–393.
71. Pollack, R., Salas, J., Wang, R., Kusano, T., and Green, H. (1971): Human-mouse hybrid cell lines and susceptibility to species-specific viruses. *J. Cell. Physiol. 77*:117–119.
72. Ptashne, M. (1967): Isolation of the lambda phage repressor. *Proc. Nat. Acad. Sci. 57*:306–313.
73. Puck, T. T., Wuthier, P., Jones, C., and Kao, F.-T. (1971): Genetics of somatic mammalian cells: Lethal antigens as genetic markers for study of human linkage groups. *Proc. Nat. Acad. Sci. 68*:3102–3106.
74. Rechsteiner, M. C. (1971): The isotopic antiglobulin assay: Application to the study of hybrid cell membranes. *J. Immunol. 107*:985–996.
75. Ruddle, F. (1972): Linkage analysis using somatic cell hybrids. *Adv. Human Genetics 3*:173–179.
76. Shevliaghyn, V. J., and Karazas, N. V. (1970): Transformation of human cells by polyoma and Rous Sarcoma viruses mediated by inactivated Sendai virus. *Int. J. Cancer 6*:234–244.
77. Smith, G. L., and Consigli, R. A. (1972): Transient inhibition of polyoma virus synthesis by Sendai virus (Parainfluenza I). I. Demonstration and nature of the inhibition by inactivated virus. *J. Virol. 10*:1091–1097.
78. Smith, G. L., and Consigli, R. A. (1972): Transient inhibition of polyoma virus synthesis by Sendai virus (Parainfluenza I.) II. Mechanism of the interference by inactivated virus. *J. Virol. 10*:1098–1108.
79. Spencer, R. A., Hauschka, T. S., Amos, D. B., and Ephrussi, B. (1964): Co-dominance of isoantigens in somatic hybrids of murine cells grown *in vitro*. *J. Nat. Cancer Inst. 33*:893–903.
80. Steplewski, Z., Knowles, B. B., and Koprowski, H. (1968): The mechanism of internuclear transmission of SV40-induced complement fixation antigen in heterokaryocytes. *Proc. Nat. Acad. Sci. 59*:769–776.
81. Steplewski, Z., and Koprowski, H. (1969): Development of SV40 coat protein antigen in nonpermissive nuclei in heterokaryocytes. *Exp. Cell Res. 57*:433–440.

82. Cited in Steplewski, Z., and Koprowski, H. (1970): Somatic cell fusion and hybridization. In: *Methods in Cancer Research*, Vol. 5, pp. 155–191 (ed. H. Busch), Academic Press, New York: Watkins, J. F. (1971): Fusion of cells for virus studies and production of cell hybrids. In: *Methods in Virology*, Vol. 5, pp. 1–32 (ed. K. Maramorosch and H. Koprowski), Academic Press, New York.
83. Suarez, H.-G., Chany, C., Cassingena, R., Vignal, M., and Lemaitre, J. (1972): Localisation du chromosome responsable de la synthèse de l'interféron simien. *C.R. Acad. Sci. Paris 274*:3622–3624.
84. Swetly, P., Barbanti-Brodano, G., Knowles, B. B., and Koprowski, H. (1969): Response of simian virus 40-transformed cell lines and cell hybrids to superinfection with simian virus 40 and its deoxyribonucleic acid. *J. Virol. 4*:348–355.
85. Takangi, N., Aya, T., Kato, H., and Sandberg, A. A. (1969): Relation of virus-induced cell fusion and chromosome pulverization to mitotic events. *J. Nat. Cancer Inst. 46*:335–347.
86. Takemoto, K. K., Todaro, G. J., and Habel, K. (1968): Recovery of SV40 virus with genetic markers of original inducing virus from SV40-transformed mouse cells. *Virology 35*:1–8.
87. Tan, Y. H., Tischfield, J., and Ruddle, F. H. (1973): The linkage of genes for the human interferon-induced antiviral protein and indophenol oxidase-B traits to chromosome G-21. *J. Exp. Med. 137*:317–330.
88. Teich, N., Lowy, D. R., Hartley, J. W., and Rowe, W. P. (1973): Studies of the mechanism of induction of infectious murine leukemia virus from AKR mouse embryo cell lines by 5-iododeoxyuridine and 5-bromodeoxyuridine. *Virology 51*:163–173.
89. Tennant, R. W., and Richter, C. B. (1972): Murine leukemia virus: Restriction in fused permissive and nonpermissive cells. *Science 178*:516–518.
90. ter Meulen, V., Iwasaki, Y., Koprowski, H., Kackell, Y. M., and Müller, D. (1972): Fusion of cultured multiple-sclerosis brain cells with indicator cells: Presence of nucleocapsids and virions and isolation of parainfluenza-type virus. *Lancet 2*:1–5.
91. ter Meulen, V., Kackell, Y., Müller, D., Katz, M., and Meyermann, R. (1972): Isolation of infectious measles virus in measles encephalitis. *Lancet 2*:1172–1175.
92. Cited in ter Meulen, V., Katz, M., and Müller, D. (1972): Subacute sclerosing panencephalitis: A review. *Current Topics Microbiol. Immunol. 57*:1–38.
93. Watkins, J. F., and Grace, D. M. (1967): Studies on the surface antigens of interspecific mammalian cell heterokaryons. *J. Cell Sci. 2*:193–204.
94. Watkins, J. F., and Chen, L. (1969): Immunization of mice against Ehrlich ascites tumour using a hamster/Ehrlich ascites tumour hybrid cell line. *Nature 223*:1018–1022.
95. Weber, J., and Stich, H. F. (1969): Response of somatic cell hybrids to exposure to adenovirus type 2 and 12. *Exp. Cell Res. 56*:319–325.
96. Weiner, L. P., Herndon, R. M., Narayan, O., Johnson, R. T., Shah, K., Rubinstein, L. J., Preziosi, T. J., and Conley, F. K. (1972): Isolation of virus related to SV40 from patients with progressive multifocal leukoencephalopathy. *N. Engl. J. Med. 286*:385–390.
97. Weiss, M. C., and Green, H. (1967): Human-mouse hybrid cell lines containing partial complements of human chromosomes and functioning human genes. *Proc. Nat. Acad. Sci. 58*:1104–1111.
98. Weiss, M. C., Ephrussi, B., and Scaletta, L. J. (1968): Loss of T-antigen from somatic hybrids between mouse cells and SV40-transformed human cells. *Proc. Nat. Acad. Sci. 59*:1132–1135.
99. Weiss, M. C. (1970): Further studies on loss of T-antigen from somatic hybrids between mouse cells and SV40-transformed human cells. *Proc. Nat. Acad. Sci. 66*:79–86.
100. Wuthier, P., Jones, C., and Puck, T. T. (1973): Surface antigen of mammalian cells as genetic markers II. *J. Exp. Med. 138*:229–244.

DISCUSSION

Davidson: Would you comment further on the possible existence of a repressor controlling or preventing the expression of latent viruses, in reference especially to the experiments involving fusion with enucleated cells? You indicated that the results suggest that there is not a repressor preventing the transformed cells from expressing viral functions. How conclusive are the results?

Koprowski: Based on the data which I have presented, we are unable to conclude that there is a repressor in SV40-transformed cells.

Watkins: How do you consider Cassingena's experiments?

Koprowski: These experiments involve crude extracts of SV40-transformed cells, and their effects are measured in terms of reduction of plaque formation by SV40. It is difficult, if not impossible, to evaluate these results in terms of a specific repressor for SV40.

Basilico: We have tried some experiments of this type. We were able to get some inhibition of plaque formation, but there was no specificity whatsoever. Also, the reduction in plaque formation is not very significant.

Somatic Cell Hybridization, edited by
Richard L. Davidson and Felix de la Cruz.
Raven Press, New York © 1974.

SV40 Rescue from a Virogenic Cell Line

John Watkins

Sir William Dunn School of Pathology, University of Oxford, South Parks Road, Oxford, England

In heterokaryons formed by fusion of SV3T3 (an SV40-transformed mouse cell) and a permissive cell, SV40 virus synthesis is initiated (3). This result raises the question of the role of the permissive cell. Does the permissive cell play an active role (for example, does it excise integrated viral DNA and provide a milieu for DNA replication and expression of all the viral functions), or does the permissive cell have only a passive role, simply providing a milieu for SV40 replication? The studies with SV3T3 suggested that the role of the permissive cell was passive because prior treatment of the nonpermissive transformed cells with IUDR increased the percentage of positive heterokaryons when such cells were subsequently fused with permissive cells (1). These observations suggest that the state of the transformed cell prior to fusion determined whether the heterokaryon would or would not form SV40 virus.

In order to study this question in more detail, I have been investigating an SV40-transformed baby rabbit kidney cell clonal line (2). All of the cells are T-antigen positive. The cells are pleomorphic and have a modal chromosome number of 44. When a culture of these cells is frozen and thawed, about 10^5 infectious units can be recovered from about 10^7 cells after 8 days of incubation. Thus there is a very slow leakage of SV40 from this transformed cell line. The cells are resistant to superinfection by intact virus. About 0.02% of the cells are V-antigen positive.

When permissive CV-1 cells were fused with the transformed baby rabbit kidney cells, heterokaryons were easily counted. Similarly, heterokaryons producing SV40 could be distinguished on the basis of morphological criteria. About 3 to 5% of heterokaryons formed in a fusion of the transformed baby rabbit kidney cells and CV-1 permissive cells produced virus. One day after fusion, 0.5% of the heterokaryons were positive, and by 3 days 3 to 5% were positive. The fraction of positive cells did not increase on further incubation of the cells. By contrast, when SV3T3 cells were fused with permissive cells, the first positive heterokaryons were detected about 4 days following fusion.

After the cells had been grown for 2 days in tritiated thymidine and autoradiographs examined, about 1 to 2% of nuclei showed a focal incorporation

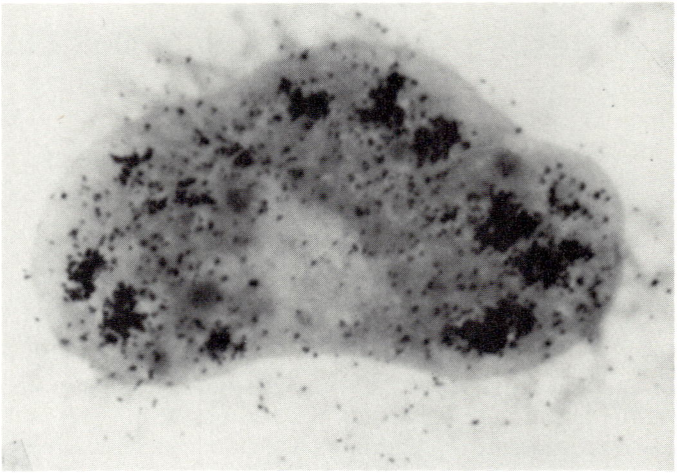

FIG. 1. Nucleus of SV40-transformed rabbit cell after 48 hr incubation in medium containing tritiated thymidine followed by autoradiography. About 1% of the nuclei showed focal incorporation of this kind.

of the isotope as shown in Fig. 1. The remainder of the cells showed diffuse nuclear labeling. When *in situ* hybridization with SV40 complementary RNA was performed by the technique of Gall and Pardue, the foci of hybridization within the cell nuclei were observed in 1 to 3% of the cells as shown in Fig. 2. About 80% of the cells showing hybridization with complementary RNA showed focal hybridization. The number of foci varied from two to 11 per nucleus, the majority of cells having four to five foci. It is tempting to speculate that each focus was the site of separate reactivation of an SV40 genome.

Neither normal rabbit cells nor the permissive CV-1 cells showed focal sites of hybridization with complementary RNA. When the permissive cells were infected with SV40 virus and examined by *in situ* hybridization 40 hr after infection, the hybridized complementary RNA was evenly distributed over the cell nuclei.

If we assumed that the V-antigen-positive cells are a subset of the cells synthesizing SV40 DNA (0.02% of the cells were V-antigen positive), it would appear that 0.7% of the cells which show *in situ* hybridization with complementary RNA were V-antigen positive. Hence 99% of the cells making SV40 DNA were V-antigen negative although the grain counts over those nuclei were at least equal to and not infrequently greater than the nuclear grain counts of SV40 infected CV-1 cells. All the permissive CV-1 cells were strongly V-antigen positive 48 hr following infection.

On the basis of these observations it is possible to make a hypothesis about the state of the virus in the transformed baby rabbit kidney cells and

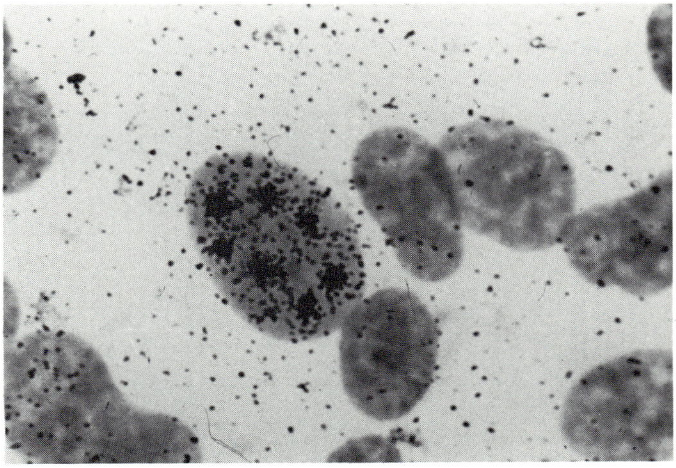

FIG. 2. Autoradiograph of a nucleus of an SV40-transformed rabbit cell showing focal distribution of *in situ* hybridization with SV40-complementary RNA. Between 1% and 3% of nuclei showed this appearance.

the role of the permissive cells. If it is assumed that the cells that showed *in situ* hybridization with complementary RNA were the same as the cells that produced SV40 virus when fused with permissive cells, and that the V-antigen-positive cells were a subset of those cells showing *in situ* hybridization, it appears that the SV40-transformed rabbit cells can exist in three states.

In the first state, state A, the SV40 DNA is not replicable. In the transformed baby rabbit kidney cells, 97% of the cells would be in state A. Cells in state A are not induced to form SV40 virus by fusion with permissive cells. State A would represent, therefore, a true latent state.

There can be a transition from state A to state B. In state B, SV40 DNA is in a form in which it is replicable and, in a cell like the rabbit cell, is in fact replicated. In mouse cells such as SV3T3, which are incapable of replicating SV40 DNA, state B would not be accompanied by replication of the viral genome. In mouse cells the virus will pass from state A to state B, where it is replicable, but will not be replicated.

Then, as a rare event, the cells can pass into a third state, state C, in which the late functions can be expressed. In heterokaryons between rabbit cells and permissive cells, all state B cells will go to state C because the permissive cells provide the milieu in which the viral DNA can be replicated and the late viral functions expressed. The permissive cells, according to this hypothesis, simply provide a milieu in which virus maturation can occur, and do not bring the viral DNA into a form in which it can be replicated.

The spontaneous transition from state B to state C does not occur in mouse cells and apparently occurs in less than 1% of the state B rabbit cells.

Since the Epstein-Barr virus in man may be equivalent to SV40 virus in rabbits, the model may also be relevant to the EB virus. State A cells, in which the SV40 genome is not expressed in a permissive milieu (when this is provided by cell fusion), may serve as a model for the nonexpression of gene functions in differentiated cells.

REFERENCES

1. Watkins, J. F. (1970): The effects of some metabolic inhibitors on the ability of SV40 virus in transformed cells to be detected by cell fusion. *J. Cell Sci. 6*:721.
2. Watkins, J. F. *(in press)*. Studies on a virogenic clone of SV40—transformed rabbit cells using cell fusion and *in situ* hybridization. *J. Gen. Virol.*
3. Watkins, J. F., and Dulbecco, R. (1967): Production of SV40 virus in heterokaryons of transformed and susceptible cells. *Proc. Nat. Acad. Sci. 58*:1396.

Somatic Cell Hybridization, edited by
Richard L. Davidson and Felix de la Cruz.
Raven Press, New York © 1974.

Fusion of Tumor Cells with Host Cells

Francis Wiener, Eva Maria Fenyö, George Klein, and Henry Harris*

Department of Tumor Biology, Karolinska Institutet, S 104 01 Stockholm 60, Sweden, and
**Sir William Dunn School of Pathology, University of Oxford, Oxford OX1 3RE, England*

There have been several claims in the literature (1, 3, 4, 7, 10, 11, 14–16) of the occurrence of cell fusion *in vivo*. However, most of these studies do not provide unequivocal evidence for the occurrence of such an event. In order to demonstrate cell fusion *in vivo* two major prerequisites must be fulfilled: first, a system is needed that permits the selection of fused cells either *in vivo* or *in vitro*, and, second, distinctive genetic markers are required to identify the hybrids in a fashion comparable to that used to define the formation of hybrid cells *in vitro*.

Our previously published experiments (2, 17) and the results presented in this volume provide evidence for the occurrence of tumor-host cell fusion in mice.

The first tumor-host cell hybrids were isolated from tumors originating from the L cell derivatives A9 and B82. The A9 cell is an 8-azaguanine-resistant derivative of the L cell line (8). It lacks the enzyme hypoxanthine guanine phosphoribosyltransferase (HGPRT) and is thus unable to grow in media such as HAT. The A9 itself has little ability to grow progressively *in vivo*. Inocula of 5×10^4 to 2×10^6 cells produced progressive tumors in only 12% of X-irradiated newborn syngeneic mice. One of these tumors was explanted and subcultured. When it was inoculated into X-irradiated newborn syngeneic or semiallogeneic mice, between 80 and 90% of the inoculated animals developed progressive tumors. The cell line was therefore designated A9HT (high-take incidence) (18). The A9HT line, like its A9 parent, lacks hypoxanthine guanine phosphoribosyltransferase and is unable to grow in HAT medium.

When tumors produced by injection of A9HT cells were explanted and grown in HAT medium, most of the cells died; however, after 8 to 14 days, clones of HAT-resistant cells appeared. These HAT-resistant clones appeared whether the tumor cells were explanted directly into HAT medium or were first grown in nonselective medium. The cells in the HAT-resistant clones were distinguishable from A9HT cells by morphological criteria. They were larger, had more fibroblastic character, and formed a variable number of extended cytoplasmic processes. Karyological investigation of these cells showed that they contained the same number of biarmed chromosomes as the A9HT, but between 30 to 40 more telocentric chromosomes. Similar experiments were done with the B82 line, an L cell derivative lacking thymidine kinase (9) and hence unable to grow in HAT. A tumor produced by B82 cells was explanted and gave rise to a B82HT line, which, like the A9HT line, produced a high-take incidence. Tumors derived from the injection of B82HT cells were explanted and cultivated in HAT; HAT-resistant colonies again arose, which proved on examination to be composed of cells with some 30 to 40 more telocentric chromosomes than the B82HT cells.

The regular appearance of these HAT-resistant cells in all the explanted tumor cell population and the fact that their chromosome constitution corresponded to what one might expect from the fusion of one A9HT or B82HT cell with one diploid mouse cell suggested the possibility that the HAT-resistant cells might be hybrids arising from the fusion of the A9HT or B82HT tumor cells with host cells. This possibility was tested by injecting the A9HT cells into animals bearing specific antigenic (H-2) or chromosomal markers (T6T6).

The A9 cell and its derivative A9HT are of C3H origin and bear the H-2^k isoantigenic complex. A9HT tumors were produced in F_1 hybrid mice (C3H,H-2^k–C57B1,H-2^b and C3H,H-2^k–ACA,H-2^f) and HAT-resistant cells were isolated from the tumors. The HAT-resistant lines were tested by quantitative absorption of specific antisera. These tests showed that the HAT-resistant lines derived from the A9HT tumors bore the H-2 isoantigen complex of the hosts in which the tumor were grown. A9HT cells were inoculated into CBA mice homozygous for the T6 translocation. Tumors produced in these animals were explanted and subjected to selection in HAT medium. Chromosome analysis of these lines revealed that they were again composed of cells having 30 to 40 telocentric chromosomes more than the A9HT cells, but, in this case, in addition to the biarmed chromosomes characteristic of the A9HT cells, the majority of the metaphase plates of these lines showed the two T6T6 marker chromosomes specific for the host cells. The karyotype of these HAT-resistant cells thus provide decisive evidence that they are hybrid cells that arose from the fusion of A9HT cells with host cells.

The question arises whether the fusion between tumor and host cells is a property confined to the enzyme-deficient A9HT or B82HT tumor cells, or whether this is a more general phenomenon. Therefore, in a second series of experiments mouse ascites tumors without selective biochemical markers (i.e., without enzyme deficiencies) were explanted after passage through appropriately marked hosts.

Obviously, the isolation of tumor-host cell hybrids in these cases would have to depend on the possibility of obtaining selective conditions that favor the hybrid cells and allow the elimination of unfused tumor cells and host cells. This was achieved by exploiting differences between the ascites tumor cells and the hybrid cells in their ability to adhere to the surface of the culture vessel.

The tumor cell lines used in these studies were SEWA and SEYF, ascitic forms of sarcomas induced by polyoma virus (12, 13) and MSWBS, an ascites sarcoma, derived from the MSWB tumor, originally induced by methylcholanthrene (6). The ascitic fluid was sucked out of the peritoneum and 0.2 ml of ascitic fluid, containing 20 to 60 $\times$ 10^6 cells, was suspended in 100 ml of medium and seeded in a Roux bottle. The medium (McCoy's 5A and MEM in equal proportion and supplemented with 10% FCS) was changed 24 and 48 hr after explantation, and thereafter every 4 to 5 days. The explanted ascites cells attached loosely to the surface of the culture vessels and could be completely removed either by shaking or trypsinization. Within 24 hr of explantation, however, a few firmly attached cells appeared as well. Some of these formed small colonies within 8 to 20 days.

These cells were completely different from the round-shaped ascitic cells. They were polygonal or fibroblast-like and tended to pile up in certain areas (Figs. 1 and 2). The hybrid nature of the lines derived from these selected colonies was established by the same criteria as those used for the A9HT hybrids. All three tumors, SEWA, SEYF, and MSWBS, were injected in F_1 hosts bearing H-2 antigenic markers different from that of the tumors and into hosts bearing the T6T6 chromosomal markers.

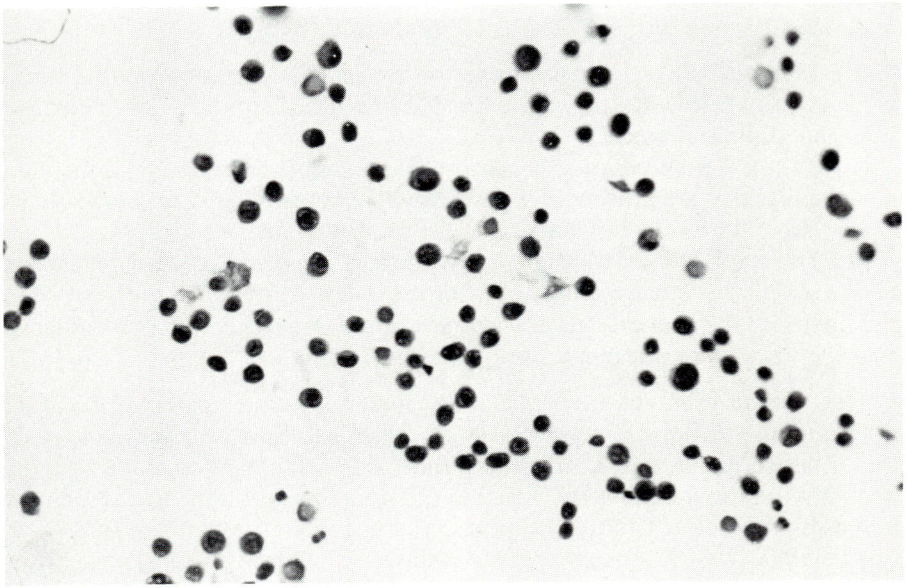

FIG. 1. Morphology of the SEWA tumor cells. H.E. $\times$ 420.

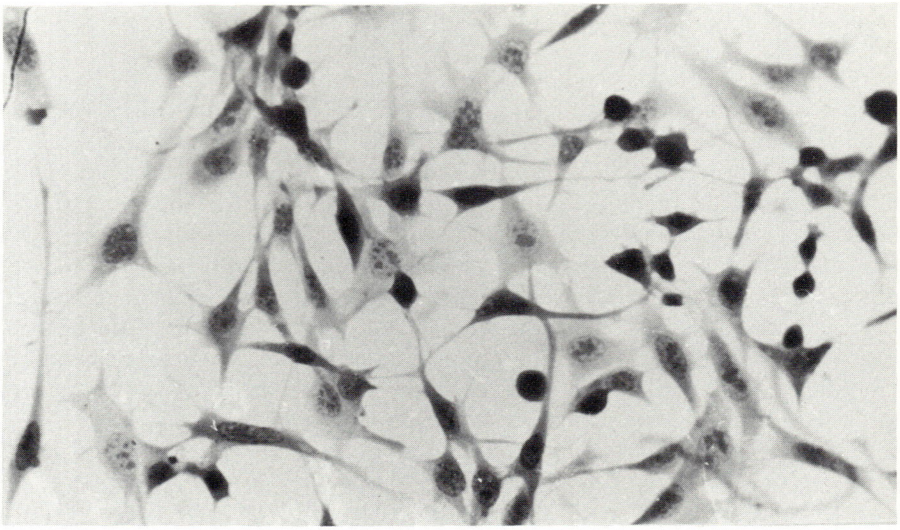

FIG. 2. SEWA-host cell hybrid (SEWA-ADA) growing as a monolayer. The population is composed of polygonal and fibroblast-like cells. H.E. × 420.

Table 1 summarizes the chromosome constitutions and the H-2 isoantigenic complexes of the ascites tumors and of the selected tumor-host cell hybrid lines. The results of the chromosome analyses and the H-2 absorption tests showed clearly that the selected lines were hybrid:

(a) The total chromosome number of the selected lines was in the range of, or slightly below, the number expected from the fusion of a tumor cell and a normal diploid host cell (Table 1).

(b) Tumor-derived marker chromosomes and host-derived T6T6 chromosomal markers were present in the expected numbers (Figs. 3 and 4, Table 1).

(c) Host H-2 antigenic markers were present (Figs. 5 and 6).

(d) The hybrid lines were tested for their malignancy on transplantation in syngeneic or semisyngeneic newborn or adult F_1 mice. Tumors arose in the majority of the inoculated animals, thus providing further evidence that the cells were not merely polyploid host cells.

The type of host cell involved in the fusion is so far unknown. In order to assess the biological significance of the phenomenon, it was obviously important to define this, at least in outline. Radiation chimeras in which the host and the repopulating cells carried different markers appeared to present a possible approach to this question.

The experimental design was as follows. F_1 hybrid mice syngeneic or semiallogeneic with respect to the test tumor were lethally irradiated and then reconstituted with CBAT6T6 hemopoietic cells. After chimerism has

TABLE 1. *Chromosome constitution of unselected and selected lines from SEWA, SEYF, and TA3 ascites tumor cells*

Unselected line							Selected lines							
		Total chromosome no.		Marker chromosomes[a]				Total chromosome no.		Marker chromosomes[a]				
Designation of cell line	Genotype of host	Range	Mode	Long	Secondary constriction (percent metaphases)	Minute	Designation	Range	Mode	Long	Secondary constriction (percent metaphase)	Minute	One biarmed	CBA T6T6
SEWA-1 (ASW)	A × ASW F_1	42–44	43	40	10	20	SEWA-ASA	75–85	81,82	40	10	20	0	0
							Tumor 1	73–83	80,83	20	10	15	0	0
							Tumor 2	57–90	77	35	5	5	0	0
							Clone 4	77–84	82,83	85	10	55	0	0
SEWA (ASW)	A × DBA/2 F_1	42–44	43	65	90	80	SEWA-ADA	76–84	81	85	40	65	0	0
							Tumor	65–86	78	90	30	40	0	0
	CBAT6T6						SEWA-BT6	77–82	80	100	90	15	100	100
				No. of biarmed chromosomes: Range		Mode					No. of biarmed chromosomes: Range		Mode	
SEYF (ABY)	A × ABY F_1	59–67	67	6–10		7	SEYF-AYA							
							Clone 10	110–135	–		4–9		6	
	ABY × C3H F_1						SEYF-YHA	97–108	104		4–9		7	
							Clone 3	82–94	84,85		5–8		6	
							Clone 1	86–101	96,97		4–7		5	
							Clone 6	95–103	99,100		5–7		5	
							Tumor	98–110	102		3–6		4	
							Clone 5	90–98	95		4–7		5	
	ABY × ASW F_1						SEYF-YSA	71–103	99		4–7		6	
MSWBS (ASW)	ASW × C3H F_1	28–30	29	8–11		10	MSWBS-SHA	89–94	91		18–22		20	
		51–57	53	18–22		20								

[a] Twenty metaphase plates were examined for each count. Fifty percent of the SEWA-1 and SEWA-ASA cells did not contain any of these three SEWA markers and 15% of SEWA and 15% of SEWA-ADA lacked all SEWA markers. One hundred percent of SEYF and MSWBS cells and the selected lines derived from them contained biarmed chromosomes.

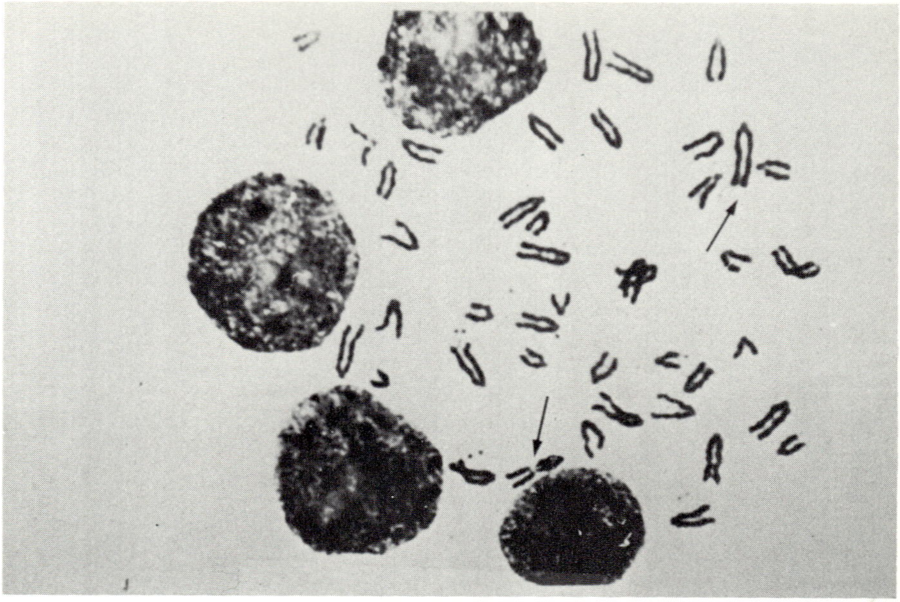

FIG. 3. Metaphase of SEWA tumor cell. The arrows indicate chromosomal markers: a long acrocentric chromosome and one with a secondary constriction. The cell contains 43 acrocentric and telocentric chromosomes.

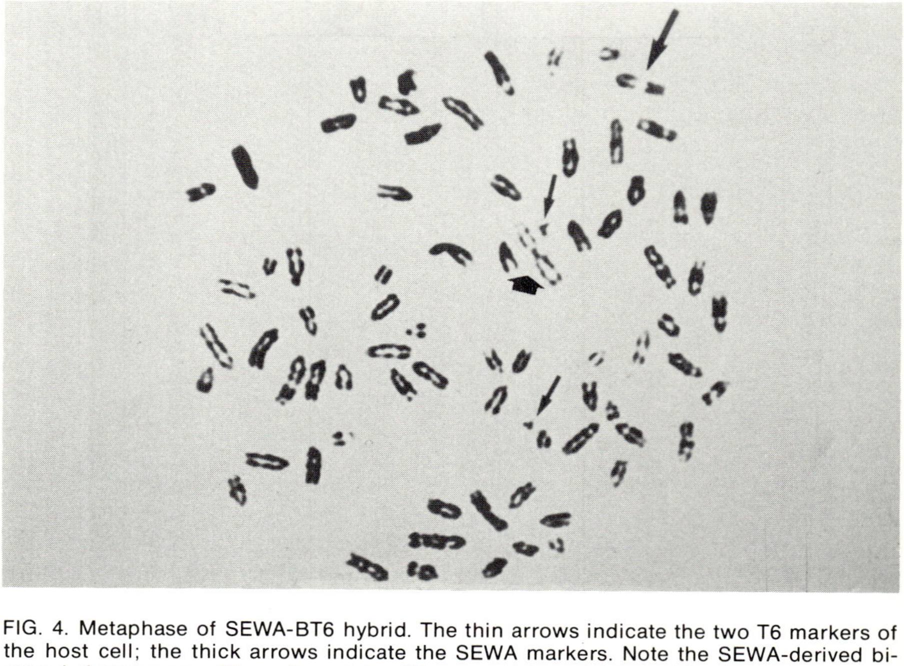

FIG. 4. Metaphase of SEWA-BT6 hybrid. The thin arrows indicate the two T6 markers of the host cell; the thick arrows indicate the SEWA markers. Note the SEWA-derived bi-armed chromosome. The cell contains 75 acrocentric and telocentric chromosomes.

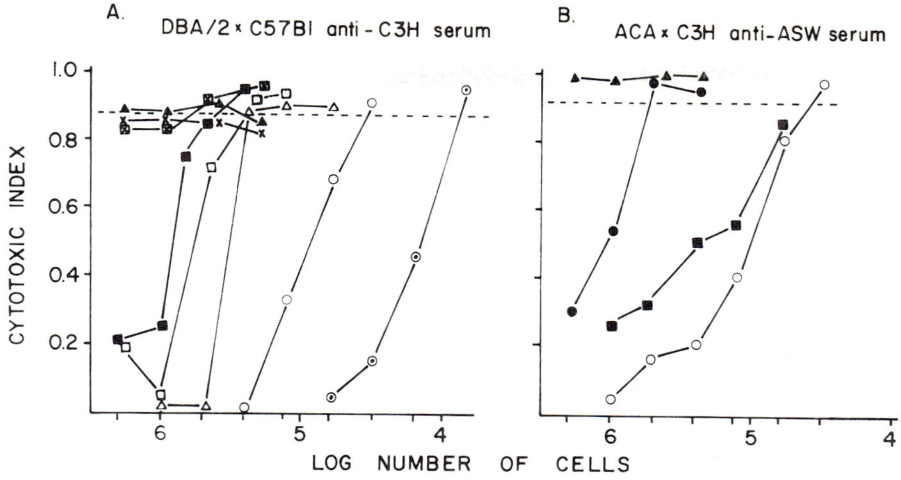

FIG. 5. Absorption of (A) DBA/2 × C57B1 anti C3H and (B) ACA × C3H anti ASW serum by the hybrid cells isolated from the SEWA, SEYF and MSWBS ascites tumors grown in A or C3H mice. × MSWBS; × SEWA; △ SEYF-AYA clone 10; ○ SEWA-ADA; □ SEYF-YHA; ● SEYF-YSA; ■ MSWBS-SHA; ⊙ SEWA-BT6D; ▲ SEYF.

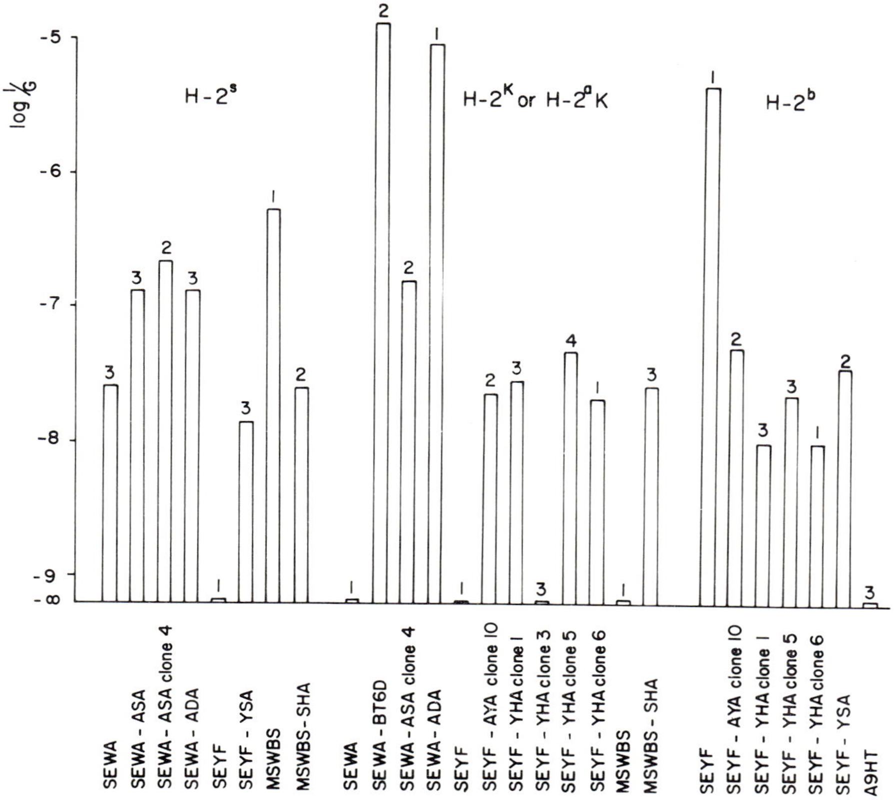

FIG. 6. Absorptive capacities of parental and hybrid lines for ACA × C3H anti ASW ($H\text{-}2^s$), DBA/2 × C57B1 anti C3H($H\text{-}2^k$ or $H\text{-}2^a K$) and ACA × C3H anti C57B1 ($H\text{-}2^b$) serum. $1/G$ indicates the amount of antibody bound per cell when 50% of the cytotoxic activity of the antiserum has been bound. The numbers at the top of columns denote the number of absorption tests done.

TABLE 2. *Chromosomes and H-2 antigens of hybrids between host cells and SEWA, SEYF, and TA3 ascites tumor cells*

Tumor	Genotype of host	Total chromosome no.		No. biarmed chrs.		T6T6 markers	H-2 Isoantigenic complex	
		Range	Mode	Range	Mode		Present	Absent
SEWA	A.SW	42–44	43	–	–	– –	$H\text{-}2^s$	
SEYF	A.BY	57–69	67	6–10	7	– –	$H\text{-}2^b$	
$TA3Bimp^-$	A	40–42	41	–	–	– –	$H\text{-}2^a$	
SEWA-H7A	C3H × C57B1 F_1	81–84	83	–	–	+ +	$H\text{-}2^s$ $H\text{-}2^k$	$H\text{-}2^b$
SEWA-H7B*	C3H × C57B1 F_1	63–86	–	–	–	+ +	$H\text{-}2^s$ $H\text{-}2^k$	$H\text{-}2^b$
SEYF-YHA	C3H × ABY F_1	82–93	–	3–7	5	+ +	$H\text{-}2^b$ $H\text{-}2^k$	
$TA3Bimp^-$-H7A*	C3H × C57B1 F_1	41–119	–	–	–	+ +	$H\text{-}2^a$	$H\text{-}2^b$
$TA3Bimp^-$-H7B	C3H × C57B1 F_1	91–122	104	–	–	+ +	$H\text{-}2^a$	$H\text{-}2^b$
$TA3Bimp^-$-H7C	C3H × C57B1 F_1	80–124	116	–	–	+ +	$H\text{-}2^a$	$H\text{-}2^b$
$TA3Bimp^-$-HDB	C3H × DBA/2 F_1	108–121	119	–	–	+ +	$H\text{-}2^a$	
$TA3Bimp^-$-HDC	C3H × DBA/2 F_1	98–120	109,115	–	–	+ +	$H\text{-}2^a$	

* Markedly reduced hybrid. The T6 markers were only present in the metaphase plates with nearly complete biparental chromosome constitution.

TABLE 3. *Chromosomes and H-2 antigens of hybrids between host cells and A9HT and B82HT solid tumors*

Tumor	Genotype of host	Total chromosome no.		No. biarmed chrs.		T6T6 markers	H-2 Isoantigenic complex	
		Range	Mode	Range	Mode		Present	Absent
A9HT	C3H	50–55	53	23–27	24	– –	$H\text{-}2^k$	
B82HT	C3H	51–56	54	22–29	25	– –	$H\text{-}2^k$	
A9HT-HDE	C3H × DBA/2 F_1	87–94	93	21–25	23	– –	$H\text{-}2^k$ $H\text{-}2^d$	
B82HT-HDA	C3H × DBA/2 F_1	82–91	86	22–30	24	– –	$H\text{-}2^k$ $H\text{-}2^d$	
A9HT-HLA	C3H × C57leaden F_1	63–95	–	15–25	–	– –	$H\text{-}2^k$ $H\text{-}2^b$	

been established, the animals were inoculated with solid or ascites tumors. Tumor-host cell hybrids were selected as previously described (2, 17). T6T6 chromosomal and H-2 antigenic markers served to distinguish between the hemopoietic cells derived from the graft and the cell of the host.

The tumors used in these studies were the A9HT and B82HT which grow in solid form and the ascites tumor cell lines SEWA, SEYF, and TA3Bimp$^-$. This last is an 8-azaguanine-resistant variant of the TA3Ha tumor, an ascitic form of a solid spontaneous mammary carcinoma (5).

All tumor-host cell hybrids isolated from the ascites tumors contained two T6T6 marker chromosomes in addition to the chromosomal and/or antigenic markers of the tumors. This shows that the hybrids were derived from the fusion of tumor cells with hemopoietic donor cells repopulating the irradiated host. The absence of host-cell participation in the case of the ascites tumors was confirmed by the absence of the host-specific H-2 antigenic markers (Table 2). For example, the SEWA-H7A-T6 hybrid was selected from a SEWA tumor (H-2^s) that grew intraperitoneally in a C3H-C57B1 F_1 (H-2^k, H-2^b) mouse, repopulated with CBAT6T6 (H-2^k) hemopoietic cells. The hybrid was H-2^s, H-2^k, T6T6+. The presence of T6T6 and H-2^k, together with the absence of H-2^b, proved that fusion has occurred between a tumor cell and a hemopoietic donor cell. Similar considerations apply to the other hybrids isolated from the three ascites tumors.

On the other hand, the hybrids selected from subcutaneous solid tumors were found to be derived from fusions between the tumor cells and the cells of the irradiated host. This was again shown by chromosomal markers and by the presence of the host-specific antigenic marker (Table 3). For example, the B82HT-HDA-T6 hybrid line is H-2^k, H-2^d T6T6−, i.e., carries the H-2^d isoantigenic complex specific for the DBA/2 host in addition to the H-2^k complex shared by the tumor and the host cell. This hybrid contains the biarmed chromosomal markers of the B82HT tumor but lacks the T6 chromosomes of the repopulating hemopoietic cells.

ACKNOWLEDGMENTS

This work was supported by Public Health Service Grant 1 R01 CA 14054–01 (N.C.I.). Grants were also received from the Swedish Cancer Society and the Jane Coffin Childs Memorial Fund for Medical Research. The work at Oxford was supported by the Cancer Research Campaign.

The expert technical assistance of Mrs. Mai-Lis Solberg, Mrs. Margareta Hagelin, and Miss Mona Hansson is gratefully acknowledged.

REFERENCES

1. Agnish, D. N., and Fedoroff, S. (1968): Tumor cell population of the Ehrlich ascites tumors. *Canad. J. Genet. Cytol. 10*:723–746.
2. Fenyö, E. M., Wiener, F., Klein, G., and Harris, H. (1973): Selection of tumor-host cell

hybrids from polyoma virus and methylcholanthrene induced ascites sarcoma. *J. Nat. Cancer Inst. 51*:1865–1877.
3. Goldenberg, M. D., Bhan, D. R., Pavia, A. R. (1971): *In vivo* human-hamster somatic cell fusion and lactate dehydrogenase profiles. *Cancer Res. 31*:1148–1152.
4. Janzen, W. H., Millman, A. P., and Thurston, O. G. (1971): Hybrid cells in solid tumors. *Cancer 27*:455–459.
5. Klein, G. (1951): Comparative studies of mouse tumors with respect to their capacity for growth as "ascites tumors" and their average nucleic acid content per cell. *Exp. Cell Res. 11*:518–573.
6. Klein, G., Klein, E. (1958): Histocompatibility changes in tumors. *J. Cell Comp. Physiol. 52*:125–168.
7. Lengerova, A. (1967): Radiation chimeras and genetics of somatic cells. *Science 155*:529–535.
8. Littlefield, J. W. (1964): Three degrees of guanylic acid-inosinic acid pyrophosphorylase deficiency in mouse fibroblasts. *Nature 203*:1142–1144.
9. Littlefield, J. W. (1966): The use of drug resistant markers to study the hybridization of mouse fibroblasts. *Exp. Cell Res. 41*:190–196.
10. Ohno, S. (1966): Cytologic and genetic evidence of somatic segregation in mammals, birds and fishes. In: Symposium on Phenotypic Expression. *In Vitro, 2*:46–60.
11. Sinha, K. (1967): Spontaneous occurrence of tetraploidy and near-haploidy in mammalian peripheral blood. *Exp. Cell Res. 47*:443–448.
12. Sjögren, H. O. (1964): Studies on specific transplantation resistance to polyoma virus induced tumors. II. Mechanism of resistance induced by polyoma virus infection. *J. Nat. Cancer Inst. 32*:375–393.
13. Sjögren, H. O., Hellström, I., and Klein, G. (1966): Transplantation of polyoma virus induced tumors in mice. *Cancer Res. 21*:329–337.
14. Stone, W. H., Friedman, J., and Fregin, A. (1964): Possible somatic cell mating in twin cattle with erythrocyte mosaicism. *Proc. Nat. Acad. Sci. 51*:1036–1043.
15. Stone, W. H., Cragle, R. G., Bendel, S., and Caulton, J. (1968): A possible second case of somatic cell mating in twin cattle with erythrocyte chimerism. *Genetics 60*:73–79.
16. Volpe, E. P., Ealey, M. E. (1970): Somatic cell mating and segregation in chimeric frogs. *Science 168*:850–852.
17. Wiener, F., Fenyö, E. M., Klein, G., and Harris, H. (1972): Fusion of tumour cells with host cells. *Nature New Biol. 238*:155–159.
18. Wiener, F., Klein, G., and Harris, H. (1973): The analysis of malignancy by cell fusion. IV. Hybrids between tumour cells and a malignant L cell derivative. *J. Cell Sci. 12*:253–261.

DISCUSSION

Davidson: Have you reinjected the hybrids that you get from the tumors to see if they are malignant?

Wiener: Yes. We have tested them for malignancy and found that all these hybrids were highly malignant.

Davidson: Have you found any tumors which do not hybridize with the host cells?

Wiener: We have tested the following tumors: A9HT, B82HT, SEWA, and MSWBS. We were able to select tumor x host cell hybrids from all of them.

Watkins: There must have been a time just after a tumor cell had fused with one of the host cells when there had been no loss of chromosomes. In order to undergo even one division, the hybrid was presumably malignant. Is this, then, a situation in which there is a complete set of tumor cell chromosomes plus a complete set of normal cell chromosomes as a result of *in vivo* fusion, and the hybrid is malignant?

Wiener: The initial hybrid cell has never been seen in the animal.

Watkins: It must have existed in the animal.

Wiener: Yes, but we can select for it only *in vitro*. There is no selective system for the hybrid *in vivo*.

Watkins: Is it possible that when you remove the tumor it contains hybrid cells with a complete set of chromosomes which are not growing in the host?

Wiener: We have never seen such a cell.

Thompson: I don't believe you can say whether the hybrid is formed *in vivo* or not. It could well be that you are removing from the animal both host and tumor cells, which subsequently hybridize *in vitro*.

Wiener: A formal proof of fusion *in vivo* is difficult to obtain. As previously reported (see ref. 17), we have explanted tumor cell populations at decreasing cell densities and found a direct relationship between the number of cells plated and the number of hybrid colonies formed. Even at densities of 100 cells per petri dish, occasional hybrid colonies are obtained. With very sparse populations of this kind, the regular production of hybrid colonies by spontaneous cell fusion *in vitro* would seem to be very improbable in the light of the reported frequency of hybrid formation between cell lines and diploid cells in other systems.

Thompson: If you had enough host cells in the *in vitro* system, you might be able to detect metaphases of the host type. You would not need a selective medium for that. If you inject the tumor and then remove the ascites fluid from the animal, it may contain a mixture of host and tumor cells that have not yet fused. If you then plated those on a nonselective medium and examined the chromosomes, you might be able to see both host and tumor cells.

Wiener: Host cells can be distinguished from the tumor cells by morphological criteria in many cases. Where we have looked for host cells in the primary explant, we have found them in very small numbers relative to the tumor cells.

Pontecorvo: Can you give an idea of the proportion of hybrid cells in the tumors?

Wiener: Approximately 10^{-5} or 10^{-6}, but that is only an estimate.

Harris: Could I try to reformulate Watkins's question? The data from hybridization of normal cells with malignant cells suggest that there is a repression of the expression of malignancy initially and that only after chromosome segregation do highly malignant lines arise [Harris, H., Miller, O., Klein, G., Worst, P., and Tachibana, T. (1969): *Nature 223*:363]. However, if the tumor cells fuse with normal cells *in vivo*, how can the hybrids begin growing? It seems to be difficult to explain how these cells begin growing, keeping in mind the repression of malignancy that the *in vitro* fusion suggests.

Wiener: But the hybrids do grow *in vivo*.

Harris: That is the point. It seems to be a dilemma.

Watkins: Possibly the hybrids are not growing *in vivo.*

Davidson: However, Wiener said that when as few as 100 cells were plated, hybrids could be found. That would suggest that the hybrid cells were growing in the tumors.

Pontecorvo: But Wiener also estimated a frequency of 10^{-5}.

Miller: It seems to me that the problem is that by the time the cells can be examined, there has already been chromosome loss.

Harris: Are you suggesting that the hybrids might lose many chromosomes at the very first division *in vivo* and proceed from that point?

Watkins: This must be the explanation.

Wiener: I think it is very probable that chromosome losses do occur at the first cell division. One might suppose that heterokaryons are formed rather frequently as the tumor grows and that these give rise, at the first division, to mononucleate hybrid cells of varying chromosome constitution resulting from mitotic irregularities of various kinds. Those mononucleate hybrids that have undergone the appropriate chromosome loss would then be selected for and would grow in the tumor. I believe that the ease and frequency with which we isolate these hybrids and their high-take incidence indicates that they are selected for and that they are growing in the animal. This is only hypothesis, however. We explant the tumor *in vitro,* and we select the hybrids *in vitro.* We do not really know what is happening *in vivo.*

Harris: If you have enough hybrid cells to pick them up when you plate out only 100 cells, whether or not there is selection *in vivo* is beside the point. There has been sufficient growth *in vivo* to produce that high a frequency when observed *in vitro.*

Minna: To give an analogous result, if HGPRT minus neuroblastoma cells are injected intracerebrally in mice, tumors are formed. If the tumors are explanted in HAT medium, you find a tremendous number of colonies. Coon and I performed these experiments. We were faced with the same dilemma that Thompson brought up, whether or not we had selected for hybrids that had formed *in vitro* rather than *in vivo.* I don't see how one can resolve that question of whether you were selecting for fusion events that occur after you put both normal cells and tumor cells together in the selective medium *in vitro.*

Ruddle: Wouldn't it be possible to culture the cells at low density on a background of feeder cells of a chromosome constitution different from the host and then examine those cells that grew up in a selective medium to see whether the hybrids had the *in vivo* marker or the *in vitro* marker? Positive results would be those where the *in vivo* marker had been recovered.

Wiener: We have injected the tumors into various organs. We have isolated hybrids between A9HT and brain cells after injecting the tumors into the brain. We have also injected tumors into liver, spleen, and different organs. We are looking for organ markers to prove that the fusion is *in vivo.*

Watkins: I would like to mention some recent work of Goldenberg et al. [Goldenberg, D., Bhan, R., and Pavia, R. (1971): *Cancer Res. 31*:1148]. They described apparent fusion *in vivo* between host cells and a human tumor growing in a hamster cheek pouch that gave rise to a transplantable hamster tumor that apparently had some human enzyme markers. This would leave no question of fusion occurring *in vitro.*

Somatic Cell Hybridization, edited by
Richard L. Davidson and Felix de la Cruz.
Raven Press, New York © 1974.

Studies of Human Alloantigens (HLA, H, P) in Man-Mouse Hybrids

Marc Fellous and Phillipe Couillin

Genetics Laboratory, Department of Biochemistry, University of Oxford, South Parks Road, Oxford, England

The results of a study on the expression of the human alloantigens, HLA, P, and H(ABO) in man-mouse somatic cell hybrids are presented in this chapter. HLA, the major histocompatibility system in man (1), is controlled by two closely linked loci, separated by 1% recombination. The system is very polymorphic and the alleles are codominately expressed. For the P system (4), five phenotypes are known: P_1, P_2, p, P_1^k and P_2^k. From family data, it would appear that P_1, P_2, and p are controlled by alleles at the same locus and are independent of the P^k phenotypes.

The expression of HLA, P, and ABO has been studied using microcomplement fixation adapted to cells in culture (3). All the anti-HLA, P, and H antisera were human alloantisera. Each specificity was determined by at least two antisera, the specificity of which were checked by microabsorption experiments using positive and negative platelets or red blood cells.

A number of man-mouse hybrid clones were obtained by fusion between the LM(TK^-) cl. 1D line derived from L cells (H-2^k) and a number of diploid human fibroblasts. The ABO, P, and HLA genotypes of the fibroblasts were established by family studies (Table 1). Fifty-three hybrid clones have been studied. Nine hybrids were from fusions with fibroblast line 106, 32 hybrids with line 126 (made by Billardon), seven hybrids with line 110, and seven hybrids with line IM. Representative results obtained with the hybrid between c1. 1D and line 110 are shown in Table 2. The H-2^k antigen was demonstrated in all clones and served as a positive control.

The expression of HLA, P, and H specificity segregates in the hybrids. The first two primary clones have only the haplotype HLA-1 and HLA-7. The other five primary clones possess the HLA antigen controlled by both the haplotypes. Upon subcloning of the line P7 A, segregation of the HLA haplotype was observed. The seven primary clones were positive for P_1 and p. Three of the six were positive for the P^k antigen. The H specificity was found only in P3 A and its two subclones. There appeared to be a correlation between HLA and P and P_1 but no correlation was observed with the other specificities.

TABLE 1. *Origin of the parental human cells used in the cell hybridizations and their genotypes for HLA, P, ABO systems*

		Antigenic phenotypes and genotypes			
		HLA[a]			
Name	Origin	1st locus	2nd locus	P1	ABO
126	Skin fibroblasts	HLA-2 W32	W18 W14	P1 homozygote (?)	A_1/O
106	Skin fibroblasts	HLA-2 W30	X HLA-5	P1 homozygote (?)	O
XM	Skin fibroblasts	HLA-3 HLA-3	HLA-7 W14	P2	A_1
110	Skin fibroblasts	HLA-1 W31	HLA-7 HLA-5	P1 homozygote (?)	O

[a] The HLA genotype of each donor was easily determined by the family studies; however, because of the known dominance of P_1 over P_2, despite the study of four to five children in each family, heterozygosity cannot be completely excluded.

TABLE 2. *c1.1D × 110 Hybrid clones. Expression of HLA (Haplotype I and II), P, H, P_k, H_2^k, on the hybrid clones and on their subclones*

Systems	HLA Haplotypes			H (ABO)	P			H_2^k
	I		II					
Line	HLA-1	HLA-7	HLA-5	H	P1	P	P^k	H_2^k
110	+++	++	++	+++	+++	++	++	—
c1.1D	—	—	—	—	—	—	—	+++
P1 N	++	++	—	—	++	++	++	++
P2 A	++	++	—	—	++	++	—	++
P3 A	++	+	++	++	++	++	++	++
Number of subclones 1	++	++	—	++	++	++	NT	++
P4 A	++	++	+	—	++	++	—	++
Number of subclones I	++	++	+	—	++	++	—	++
2	—	—	—	—	—	—	—	++
1	++	++	—	—	++	++	—	++
P5 A	++	++	++	—	++	++	—	++
P6 N	+	+	+	—	++	+	NT	++
P7 A	++	++	++	—	++	++	++	++
Number of subclones 1	—	—	—	—	—	—	NT	++
1	++	++	++	—	++	++	++	++
4	—	—	++	—	NT	++	NT	++
2	++	++	—	—	NT	++	NT	++

+ weakly positive reaction; +++ strongly positive reaction; — negative reaction; ++ positive reaction; NT nontested.

Similar studies have been performed on a number of other hybrids and the results are summarized in Table 3. There is a positive correlation between the expression of HLA and p or P_1 specificities. There was only one discordant clone. There was no apparent correlation between HLA and P^k nor between the HLA and H specificities.

At least two hypotheses could explain these observations. If the loss of human chromosomes from the hybrid cells is random, the HLA and P loci could be syntenic. However, antigenic markers are complex structures. It is therefore also possible that the HLA and P antigenic determinants could be expressed on a unique structure or molecule on the cell surface membrane. Some evidence that supports the first hypothesis, i.e., the synteny of HLA and P, has been obtained.

To determine whether the HLA and P antigens are expressed on independent molecules or on one complex molecule, experiments using the antibody-induced redistribution phenomena were performed with the collaboration of C. Sautés. It has recently been shown that at 37°C bivalent antibodies induce at the surface of lymphocytes a redistribution of histocompatibility antigens that cluster in spots. These clusters are visible when

TABLE 3. *Study of the correlation between the presence and the absence of antigens of the HLA system and P, P1, in the man-mouse hybrid clones studied*

P, and/ or P1 expression		HLA expression +	HLA expression −
	+	34	1
	−	0	18

In the 53 independent hybrid clones tested, HLA and P segregate simultaneously.

P^k expression		HLA expression +	HLA expression −
	+	16	1
	−	16	18

On the contrary there is no positive correlation between HLA and P^k expression.

H (ABO) expression		P or HLA expression +	P or HLA expression −
	+	17	10
	−	17	9

There is no correlation between HLA or P and H expression.

the cells are examined with immunofluorescence. If HLA and P antigens are expressed on the same structure, the redistribution of the P determinant induced by anti-P antibodies should also induce a redistribution of the HLA determinant. However, most of the doubly labeled cells have an aggregated distribution of the first conjugate, which is the P specificity, and a diffuse distribution of the second conjugate. These data, therefore, suggest that HLA and P may be located on different molecules.

An additional argument for the synteny of HLA and P would be family data. There is, however, only weak evidence for loose linkage between HLA and P with a distance of 0.30 and a lod score of 1.24 (2).

It has been suggested that there is a relationship between the product of the H-2 genes and the presence of a viral receptor. Therefore, we have studied the sensitivity to echovirus and poliovirus type 1 of somatic cell hybrids isolated from a fusion of c1.1D and the fibroblast line 110. Both viruses contain single-stranded RNA, are cytotoxic, and belong to the same family of human papova virus. These two viruses were chosen because the receptors have been well defined and because the mouse parental cell is also resistant to these viruses. As shown in Table 4 some of the hybrid clones are susceptible to the viruses and some are not. There appeared to be a correlation between the sensitivity of the clones to the two viruses, which could mean that there is a common receptor or that the genes that control these receptors are linked. However, when the susceptibility of the clones to the viruses was compared to the expression of HLA, three clones (P2B, P4B, P4A), which were HLA minus, did produce infectious virus particles (Table 4). These results suggest that HLA does not serve as a membrane receptor for at least these two viruses.

I would like to emphasize some of the difficulties encountered in the study of alloantigens in hybrid cells. There is some phenotypic variation in the expression of some antigens perhaps due to antigenic masking or antigenic suppression. These phenomena are responsible for false negative results. The expression of the antigens in the hybrids is not infrequently weaker by a factor of four to five than in the human parental cell. False positive results could also be obtained due to the presence of heteroantibodies against mouse antigens in all of the human sera used, and they always need to be absorbed with the mouse parental cell before use. Complex antigens like ABO or P have a multistep biosynthesis controlled by independent genes. The absence of an antigen could be due to the loss of the gene determining the synthesis of one of the biosynthetic steps. Our studies in man-mouse hybrids have provided some information on the biosynthesis of the complex antigen P. Consider a theoretical complex antigen C, for which either A or B is the precursor. If A is the precursor of B, then in hybrid clones it would be impossible to observe the presence of B in the absence of A. If B were the precursor of A, the reverse would be true. The correlation between the antigens p or P_1 and P^k is shown in Table 5. Based on these observations we have proposed

TABLE 4. *Sensitivity of hybrid clones 110 × c1.1D to poliovirus (type 1) and echovirus (type 11). Study of correlation with the expression of HLA antigen*

Virus	Hybrid clones											Parental cells		Sensitive cells used for titration	
	P1 N	P2 A	P2 B*	P3 A	P4 A	P4 B*	P4 N*	P5 A	P5 N	P7 A	P7 b	Human 110	Mouse c1.1D	Human embryonic cells L 809	Kidney cells Vero
Echovirus titration in infectious particle/ml	0 (2)	10^2 (6)	10^3 (2)	10^4 (4)	10^3 (1)	5.10^5 (2)	10^6 (5)	0 (5)	0 (3)	0 (3)	0 (3)	10^5 (3)	0 (10)	10^7 (10)	NT
Poliovirus titration in infectious particles	0 (2)	10^4 (2)	10^4 (2)	10^5 (2)	10^7 (2)	10^4 (2)	10^6 (2)	0 (2)	0 (2)	0 (2)	0 (2)	10^6 (2)	0 (2)	10^7 (2)	10^6 (2)
HLA expression	+	+	−	+	+	−	−	−	−	+	−				

The echovirus 11 and poliovirus type 1 belong to the same family of human picornaviruses. They are single-stranded RNA viruses, which have been chosen because the mouse parental line c1.1D is resistant to them.

Scheme of the experiment with poliovirus: 10^5 attached cells are infected with poliovirus at an infection multiplicity of 1 and 0.1. After 1 hr at 37°C, the cells are washed three times using a neutralizing antisera. The cells are incubated for 10 and 24 hr. Then the cells are disrupted by freezing and thawing. The cell debris is separated by centrifugation for 10 min at 8000 rpm. The titration is made on Vero cells (monkey kidney cells).

The scheme of the experiment with the echovirus is similar but the infectivity of the supernatant of the culture is titrated on a human embryonic line (L 809).

TABLE 5. *Study of the correlation between the presence or absence of the antigens of P, P_1, and P^k system in man-mouse hybrid clones*

		P P_1 +	P P_1 −
P^K	+	16	0
	−	16	19

Results expected according to Kortekangas, Sanger, Race, hypothesis (P^k precursor of PP_1)

		PP_1 +	PP_1 −
P^K	+	N1	N2
	−	0	N3

Necessary hypotheses—The genes which control these biosynthetic sequences are independent from one another. There is never a complete transformation of the basic substrate.

that in the biosynthetic pathway of the P system, the action of the P^k genes occurs after that of the P_1 or P_2 gene. This hypothesis is compatible with data obtained from the study of rare phenotypes (p) of the P system.

We have recently attempted to assign the HLA locus to one of the human chromosomes. One hybrid clone, isolated in HAT (by Dr. Hors) contained only one human chromosome and was HLA positive. A subclone isolated in BUDR was HLA negative. Twenty-one additional clones have been analyzed. One discordant clone was observed in HAT medium. In BUDR all the clones are HLA negative. Three hypotheses can be suggested for the BUDR results. (a) The expression of HLA was suppressed in BUDR. However, HLA was not re-expressed when the cells were grown in the absence of BUDR. (b) The loss of human chromosomes induced by BUDR is not limited to the human chromosome number 17. (c) The BUDR resistance involved two events: the loss of thymidine kinase and also the loss of an undefined gene controlling the uptake of BUDR by the cell. We could then be studying the segregation of these two genes.

REFERENCES

1. Dausset, J., Colombani, J., Legrand, L., and Fellous, M. (1970): Genetics of the HLA system. *Histocompatibility Testing,* Munksgaard, Copenhagen, p. 59.
2. Edward, J. H. (1972): Appendix of the joint report of the 5th International Workshop on Histocompatibility. *Histocompatibility Testing,* Munksgaard, Copenhagen.
3. Fellous, M., Couillin, P., Neauport-Sautes, C., Frezal, J., Billardon, C., and Dausset, J. *(in press):* Studies of human alloantigens on man-mouse hybrids. Possible syntheny between HLA and P system. *Eur. J. Immunol.*
4. Race, R. R., and Sanger, R. (1968): *Blood Groups in Man.* Blackwell, London, p. 136.

Somatic Cell Hybridization, edited by
Richard L. Davidson and Felix de la Cruz.
Raven Press, New York © 1974.

Surface Antigen Mobility in Heterokaryons

Michael Edidin

Mergenthaler Laboratory for Biology, The Johns Hopkins University, Baltimore, Maryland 21218

Our work on the regulation of membranes in early heterokaryons will be discussed. Experiments by Dr. Watkins (8), by Dr. Gordon (4), and by my own laboratory (3) indicated that a variety of cell surface markers, including antigens and enzymes, rapidly redistributed in the membrane of newly formed heterokaryons. The kinetics of redistribution were studied and we attempted to demonstrate that redistribution occurred by simple diffusion rather than by any other process.

Since then a number of lines of evidence have supported the concept of a fluid membrane. Electron-spin resonance studies have shown translational motion of spin-labeled molecules and have also provided a measure of the diffusion constants (6).

Fluorescence polarization studies of membrane viscosity have also been carried out (5). Thus, there are a number of measurements that show either that the membrane lipid is in a fluid-like state or that surface-membrane proteins are free to move. The two types of experiments are quite different. The lipid measurements are measuring what may be called fluidity, while experiments on the migration of membrane proteins measure protein translational mobility. The latter implies a fluid membrane but may make further demands on the system. We have recently measured the diffusion constants of antigens in cultured myotubes (2). The data agrees well with previous measurements of diffusion constants.

Another type of experiment that implies a fluid membrane and protein translational mobility in the plane of the membrane is the formation of "caps" in a variety of cells. If certain cells are reacted either with anti-immunoglobulin or with antisera to another cell surface antigen and are incubated at 37°C for about 60 min, the stain is no longer circumferential, but is aggregated at one pole of the cell (7). We became interested in cap formation in fibroblasts and have been studying this. The results suggested that, for cap formation, it is necessary to have not only the movement of antigens to form cross-linked lattice of antigens and antibody, but also some type of net membrane flow, perhaps of the type described by Abercrombie (1) for cultured fibroblasts. This seemed to involve a sweep of the membrane back from the ruffled leaking edge of the fibroblast to "an unknown sink."

Our previous capping studies had been done on L cells, c1.1D, which have small, neat, unspectacular ruffles. There are great morphological differences between the rather restricted ruffles of c1.1D and the exuberant ruffles of primary fibroblasts. It was assumed that cells with good ruffles should cap rapidly. To study cap formation in cells with marked ruffling, we prepared mouse fibroblasts and also studied BALB/c 3T3 and human WI-38 cells. However, none of them showed any rearrangement of the stain, neither capping nor speckling, which would have been expected if the antigens were free to move.

We considered the possibility that cells with good ruffles might not show cap formation because their membrane proteins might be restricted in their motion. We decided to utilize heterokaryons to study this possibility. It should be noted that the cells used to show rearrangement of antigens in fusion systems or in capping studies are either cells like macrophages or lymphocytes, which are normally free living, or are cells that are malignant or are not contact inhibited for growth or movement. We determined what happens when contact inhibited cells are fused and the heterokaryons are examined for the position of surface antigens at various times after fusion. The mouse antigens studied were the H-2 antigens, while for the human cells, we used either an HL-A antigen or the human species antigens. The data is presented in Table 1. For each number in the columns, about 40 cells/sample/time point were classified and averaged. The best data is from the Balb 3T3 by WI-38 cross which has been repeated five times. Most of the other numbers are the average of two experiments. In this sort of experiment, differences should be at least 1.5-fold to be significant unless the number of experiments is quite large.

In the clone 1D by VA-2 cross, a fraction of the population shows segregation of the mouse and human antigens at 60 min after fusion. By 2 hr, essentially all of the cells show intermixing of the two antigens. In the BALB 3T3 by WI-38 cross, about 90% of the cells show segregation at 1 to 2 min. The fraction of segregants decreases for about 60 min and then appears to

TABLE 1. *Segregation of human and mouse surface antigens*

Cell pair	Percent segregates of human and mouse surface antigens				
	10	60	120 minutes	180	240
Balb 3T3–WI38	68	44	43	27	31
3T3–VA2	75	58	31	–	–
FRSV 3T3–VA2	80	60	50	–	–
SV 3T3–VA2	50	10	8		
c1.1D–VA2	71	25	5		
3T3–2° human	93	79	80		

level off. The 3-hr and 4-hr time points have been studied only once. We could be observing a slow intermixing rate of the two antigens, or there could be a constant fraction of cells that does not show intermixing. In any case, many more segregated cells persist in the 3T3–WI-38 heterokaryons than in the clone 1D–Va-2 heterokaryons. Heterokaryons formed between 3T3 and VA-2 appear to be similar to the cl.1D–VA-2 heterokaryons. We do not have much data on the SV40-3T3 by WI-38 cross yet but the cells appear to behave similarly to the 3T3 by WI-38 heterokaryons.

Heterokaryons produced by fusion of SV40-3T3 and VA-2 behave similarly to the clone 1D–VA-2 heterokaryons. There is a rather rapid intermixing of the two antigens such that by 1 to 2 hr after fusion essentially no segregated cells remain. We have also tested some "flat" revertants of the SV40-3T3 cells. When fused with VA-2 the heterokaryons are similar to those of the 3T3–VA-2 cross showing a high proportion of segregated cells at 10 min and a subsequent fall off by 60 min. It is not yet clear what happens after 2 hr in this cross.

The heterokaryons that show the slowest mixing are those from a cross of 3T3 by secondary human kidney fibroblasts. There is an initial slow decline in the fraction of segregated cells but at 2 hr about 80% of the cells are still segregants.

On the basis of our data it would appear that in crosses of contact inhibited by contact inhibited cells or contact inhibited by transformed cells the fraction of segregant cells slowly approaches an asymptote or else the two sets of antigens are redistributed very slowly. When SV40-transformed cells are crossed with SV40-transformed cells, the fraction of segregants falls very rapidly as in our original heterokaryons between clone 1D and VA-2. It would appear that there is a definite difference in the intermixing behavior of the heterokaryons. Our observations must, of course, be extended to other cell pairs which can be defined as contact inhibited or noncontact inhibited. We are also interested in understanding why the restriction of intermixing of the two antigens occurs. By growing the cl. 1D–VA-2 heterokaryons in medium with the same high serum content used for the 3T3–WI-38 cells, we were able to show that the serum concentration has no effect on the intermixing rates of the two antigens.

We do not know whether restriction of intermixing reflects a modification of the membrane lipids, or of the mobility of the proteins. We are planning to probe these possibilities with fluorescence polarization studies.

REFERENCES

1. Abercrombie, M., Heaysman, J. E. M., and Pegrum, S. (1970): The locomotion of fibroblasts in culture. III. Movements of particles on the dorsal surface of the leading lamella. *Exp. Cell Res.* *62*:389–398.

2. Edidin, M., and Fambrough, D. (1973): Fluidity of the surface of cultured muscle fibers. *J. Cell Biol.* *57*:27–37.
3. Frye, L. D., and Edidin, M. (1970): The rapid intermixing of cell surface antigens after formation of mouse-human heterokaryons. *J. Cell Sci.* *7*:319–335.
4. Gordon, S., and Cohn, Z. (1970): Macrophage-melanocyte heterokaryons. I. Preparation and properties. *J. Exp. Med.* *131*:981–1003.
5. Rudy, B., and Gitler, C. (1972): Microviscosity of the cell membrane. *Biochim. Biophys. Acta* *288*:231–236.
6. Scandella, C. J., Devaux, P., and McConnell, H. M. (1972): Rapid lateral diffusion of phospholipids in rabbit sarcoplasmic reticulum. *Proc. Nat. Acad. Sci. USA* *69*:2056–2060.
7. Taylor, R. B., Duffus, W. P. H., Raff, M. C., and dePetris, S. (1971): Redistribution and pinocytosis of lymphocyte surface immunoglobulin molecules induced by anti-immunoglobulin antibody. *Nature New Biol.* *233*:225–229.
8. Watkins, J. F., and Grace, D. M. (1967): Studies on the surface antigens of interspecific mammalian cell heterokaryons. *J. Cell Sci.* *2*:193–204.

DISCUSSION

Ruddle: What were the conditions under which you carried out these experiments? Are the cells fused in suspension or attached to the surface? How long are the cells in contact with Sendai virus?

Edidin: The cells are fused in suspension. The cells are agglutinated in the cold for 10 min at pH 7.6. The cells are then shaken at 37°C and the time points are taken from the transfer to 37°C.

We have done controls in which the reaction was stopped by chilling the mixture on ice. Some cells were stained immediately and some were held overnight. The reaction really seems to be arrested by chilling.

Somatic Cell Hybridization, edited by
Richard L. Davidson and Felix de la Cruz.
Raven Press, New York © 1974.

Control of Expression of Differentiated Functions in Somatic Cell Hybrids

Richard L. Davidson

Genetics Division, Children's Hospital Medical Center and Department of Microbiology and Molecular Genetics, Harvard Medical School, Boston, Massachusetts

INTRODUCTION

It is generally assumed that all of the cells of a higher organism contain the same genetic information and that differentiation is based on the selective expression of different parts of that information. However, the mechanisms that function to control selective gene expression in differentiated cells are not well understood.

One way of studying these mechanisms of gene regulation involves combining into a single nucleus the genomes of two cells that are phenotypically different. With the techniques of somatic cell hybridization, cells that are expressing a differentiated function can be fused with cells that do not express the function. The resulting mononucleate hybrid cells seem to be capable of unlimited proliferation and can be analyzed biochemically and cytologically. (For a general review of cell hybridization, see ref. 9.) It is assumed that both parental cells have the genetic information necessary for the expression of the differentiated function, but that only one of the parental cells is actually expressing it. The hybrid cells are analyzed in order to obtain evidence on the mechanisms responsible for the difference in gene expression.

Generally the hybridization experiments have involved the fusion of the cells of different species, e.g., hamster-mouse or human-mouse. This makes it possible to monitor the activity of a large number of genes of the two parental cells in the hybrids, since homologous enzymes from different species can often be resolved on the basis of their physicochemical properties (e.g., lactate dehydrogenase, malate dehydrogenase, glucose 6 phosphate dehydrogenase, etc.). On the other hand, the use of the cells of different species as parental cells introduces some obvious difficulties into the interpretation of the results of the experiments, since some changes in gene expression in the hybrids could possibly be due to the combination of the genomes of different species and not to mechanisms of gene regulation.

All of the experiments on gene regulation in hybrid cells have involved heteroploid cells of permanent lines, rather than normal diploid cells. It is assumed that the results obtained with the hybrids can be extrapolated to the processes of differentiation in normal cells. At present, this remains an assumption that must be tested.

In a hybrid resulting from the fusion of a cell that expresses a differentiated function with a cell that does not, three types of results could be expected a priori. (a) The hybrids do not express the differentiated function. (b) The hybrids do express the function, with the relevant genes of the differentiated parent continuing to be active in the hybrids, whereas the corresponding genes of the undifferentiated parent remain unexpressed. (c) The hybrids express the function, with the relevant genes of the undifferentiated parent being expressed in addition to those of the differentiated parent.

The first result is consistent with the existence of a soluble regulator substance that controls a differentiated function by "suppressing" its expression. The second result provides no evidence of interaction between the genomes in the hybrid cells. The third result is consistent with the existence of a soluble regulator substance that controls a differentiated function by "activating" its expression. It should be emphasized that the experiments give results that are *consistent* with various mechanisms but, in themselves, are not proof of the existence of such mechanisms. Clearly other mechanisms consistent with the different types of results could also be proposed. It should also be kept in mind that the use of the term "suppression" is not meant to imply any specific mechanism of regulation. Rather it is simply a way of indicating the disappearance of a function in a hybrid cell.

REVIEW

The mechanisms controlling the expression of a variety of differentiated functions have been analyzed in experiments involving the hybridization of differentiated cells with undifferentiated cells (i.e., cells that do not express the differentiated function). The results are summarized in Table 1. In the first experiments of this type, pigmented Syrian hamster melanoma cells were hybridized with mouse fibroblasts. In the hybrids, pigment synthesis was always absent (7, 11–13). Similar results were obtained with hybrids between mouse melanoma cells and mouse fibroblasts (28). Recently it was shown that pigment synthesis could be maintained in the hybrids by doubling the ratio of pigment cell: fibroblast genomes (8, 15) (see below).

The expression of the multiple functions characteristic of glial cells (synthesis of the protein S100, high levels of the enzyme glycerol phosphate dehydrogenase, and the inducibility of the enzyme by hydrocortisone) was studied in hybrids between rat glioma cells and mouse fibroblasts that expressed none of those functions. All of the functions were greatly reduced or completely absent in the hybrids (1, 10). These experiments were the first in which changes in gene dosage were used to study the regulation of dif-

TABLE 1. *Fate of differentiated functions in hybrid cells*

Function[a]	Cross[b]	Comments[c]	Ref.
(1) Loss of Activity in Hybrids			
Melanin	SH or M melanoma x M	Dosage effect	7, 8, 11, 12, 13, 15, 28
S100 protein	R glioma x M		1
GPDH and inducibility	R glioma x M	Dosage effect	10
Growth hormone	M pituitary tumor x M		29
ES-2	M kidney tumor x M or H	Segregation	18
Liver ADH	R hepatoma x M or R	Segregation	3
Aldolase B	R hepatoma x R	Segregation	4
TAT and inducibility	R hepatoma x R or M	Segregation	2, 27, 31
AAT and inducibility	R hepatoma x R	Segregation	30
TP and inducibility	M liver x M		26
Glycogen accumulation	M liver x M		26
Steroid sulfatase	M neuroblastoma x M		19
Immunoglobulin	M myeloma x M		5, 22, 23, 24
Multipotency	M teratoma x M		14, 16
(2) Retention of Activity in Hybrids			
Excitable membranes	M neuroblastoma x M		20, 21
Acetyl cholinesterase	M neuroblastoma x M		20, 21
14–3–2 protein	M neuroblastoma x M		19
Albumin	R hepatoma x M	Dosage effect	25
Kappa light chain	M myeloma x M		22, 23
AHCH and inducibility	M fibroblast x M or R		2, 33
(3) Appearance of New Activity in Hybrids			
C4 (human)	GP macrophage ($C4^-$) x H		6
Esterases I, II, III (hamster)	CH fibroblast x H		17

[a] Abbreviations for differentiated functions: GPDH-glycerol phosphate dehydrogenase; ADH-alcohol dehydrogenase; TAT-tyrosine aminotransferase; AAT-alanine aminotransferase; TP-tryptophan pyrrolase; AHCH-acyl hydrocarbon hydroxylase; and C4-fourth component of complement.

[b] For each cross, the cell lines to the left of the x are the lines which express the differentiated function; those to the right of the x do not express the function. The species of the cells are indicated by the letters M-mouse, SH-Syrian hamster, R-rat, H-human, GP-guinea pig, and CH-Chinese hamster.

[c] The comments indicate if a given result is altered by a change in genome ratios or by the loss of chromosomes.

ferentiated functions, and the results provided information on the coordination of multiple functions (see below).

Growth-hormone synthesis was analyzed in hybrids between rat pituitary tumor cells and mouse fibroblasts, and the differentiated function was found to be absent in the hybrids (29). The kidney-specific esterase ES-2 was analyzed in hybrids between mouse renal adenocarcinoma cells and mouse or human fibroblasts (18). The kidney-specific enzyme was absent from the hybrids, but, following the loss of human chromosomes in the human-mouse hybrids, ES-2 activity reappeared (see below).

The expression of a number of differentiated functions has been studied in hybrids between liver cells, derived either from rat hepatoma or mouse fetal liver, and mouse or rat fibroblasts. The majority of the differentiated characteristics (liver alcohol dehydrogenase, aldolase B, glycogen accumulation, and high levels of tyrosine aminotransferase, alanine amino transferase, and tryptophan pyrrolase, and the inducibility of these enzymes by hydrocortisone) were absent from the hybrids (2–4, 26, 27, 30). Upon chromosome loss, however, some of these functions reappeared, and the patterns of reappearance provided further information on the coordination of multiple functions (3, 4, 30, 31) (see below). One function characteristic of liver cells, albumin synthesis, was retained in hybrids between rat hepatoma cells and mouse fibroblasts (25). By changing the ratio of hepatoma: fibroblast genomes in the hybrids, it was possible to obtain hybrids that were producing not only the rat (hepatoma) albumin but also the mouse (fibroblast) albumin (see below).

Several nerve-specific functions have been studied in hybrids between mouse neuroblastoma cells and mouse fibroblasts. One enzyme characteristic of the neuroblastoma cells, steroid sulfatase, was absent from the hybrids (19). However, most of the nerve-specific functions (excitable membranes, acetyl cholinesterase, and protein 14–3–2) were retained in the hybrids (19–21). It was noted that some of these functions could be present in the hybrids although others were absent. Immunoglobulin production has been studied in hybrids between mouse myeloma cells and mouse fibroblasts. In the hybrids, immunoglobulin production was generally absent (5, 22–24). However, the secretion of kappa light chains was observed (22, 23), and, in a few cases, immunoglobulin production was also detected (22).

Experiments have also been performed on the *potential* of cells to differentiate. Teratoma cells that, during long proliferation *in vitro,* retained the ability to form several differentiated tissues when reimplanted into mice were hybridized with mouse fibroblasts. When the hybrids were injected into mice, tumors that contained no differentiated elements developed (14, 16).

A few examples were cited above in which differentiated functions were maintained in hybrids. Another example involves the production of aryl hydrocarbon hydroxylase and its inducibility by benzanthracene. High levels of enzyme activity and inducibility were retained in hybrids produced by fusing cells of different mouse fibroblast lines with each other (33). In hybrids resulting from the fusion of cells of different Syrian hamster fibroblast lines with each other, enzyme inducibility was retained, even though the noninduced level of enzyme had dropped to undetectable levels (33).

In a few cases, functions have been observed in hybrids between cells neither of which expressed the function. Macrophages derived from guinea pigs genetically unable to produce the fourth component of complement (C4) were hybridized with human cells (HeLa). C4 production occurred in the hybrids, and it was shown that the C4 was of the human type (6).

In another experiment, hybrids between Chinese hamster ovary cells and human fibroblasts were found to produce three new esterases. These new enzymes were found to correspond to three esterases that occur in Chinese hamster ovary but were not produced by the hamster cells in culture (17). The production of the three hamster esterases in the hybrids seemed to depend upon the presence of a specific human chromosome.

Regulation of Pigment Synthesis in Hybrid Cells

Having briefly described the hybrids with differentiated cells that have been isolated thus far, a few systems, which seem best to illustrate the features of regulation that have been observed in hybrid cells, will now be discussed in detail. The regulation of melanin synthesis has been extensively studied in hybrids between pigmented Syrian hamster melanoma cells (3460–3) and mouse fibroblasts. Since the unpigmented fibroblasts had been ultimately derived from pigmented mice, it was assumed that these cells contained the genes necessary for melanin synthesis but were not expressing them.

Within a few generations after the formation of the hybrid cells, it became clear that the hybrids were unpigmented (12, 13). More than 100 independently derived hybrids (from crosses between the hamster melanoma cells and cells of three different mouse fibroblast lines) have been observed and all were unpigmented (see Fig. 1). The hybrids remained unpigmented during prolonged periods of growth.

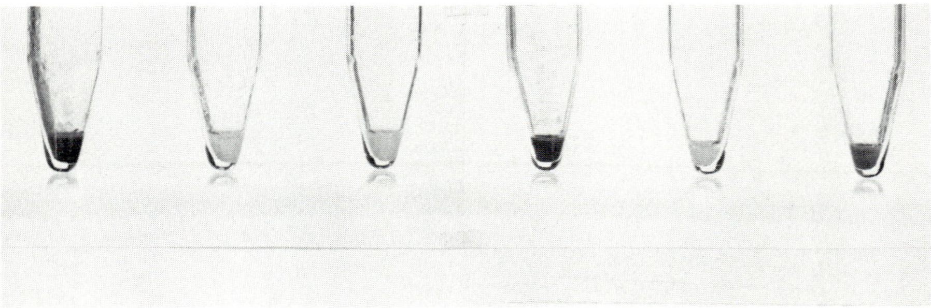

FIG. 1. Phenotypes of melanoma–fibroblast hybrids. Pelleted living cells of parental and hybrid lines. From left: (1) Syrian hamster melanoma 3460–3; (2) mouse fibroblast; (3) 3460–3 x fibroblast hybrid; (4) near tetraploid melanoma $3460–3^2$; (5) unpigmented $3460–3^2$ x fibroblast hybrid; (6) pigmented $3460–3^2$ x fibroblast hybrid.

Studies of the enzyme dopa oxidase, the only enzyme necessary for the conversion of tyrosine to melanin, were undertaken in order to obtain more specific information about the regulation of pigment synthesis. It was found that the hybrid cells, like the fibroblasts, had no detectable dopa oxidase activity, whereas the melanoma cells had very high activity. In fact, the unpigmented cells even had an inhibitor of dopa oxidase activity. However, kinetic experiments indicated that this inhibitor was not the reason for the absence of enzyme activity in the unpigmented cells (11). Kinetic experiments further provided no evidence for interactions at the level of enzyme activation or enzyme stabilization. Since there was no evidence for regulation at any other level, it was suggested that the enzyme dopa oxidase was not synthesized in hybrids between pigmented and unpigmented cells.

Before the absence of pigment synthesis in the hybrids could be taken as evidence for the operation of a regulatory system, other possibilities had to be ruled out. It had to be shown that the effect on pigment synthesis was specific, and did not represent a nonspecific suppression of the melanoma genome in the hybrids. In order to test this, the hybrids were examined for a group of enzymes not related to pigment cell differentiation and for which the forms synthesized by the hamster and mouse parental cells could be resolved (13). In all cases, the forms of both parental cells were present in the hybrids. This is shown for the enzyme malate dehydrogenase in Fig. 2. It was thus clear that the entire melanoma genome was not suppressed in the hybrids.

It also had to be shown that the melanoma population was homogeneously pigmented, i.e., that the melanoma cells that fused with the fibroblasts to give unpigmented hybrids were actually cells capable of pigment synthesis. Using Sendai virus to induce cell fusion, it was possible to get one out of every 25 melanoma cells to fuse and form a viable hybrid (7). Since all of the hybrids were unpigmented, whereas the frequency of amelanotic variants in the melanoma population was less than one in 30,000, it was clear that the melanoma cells which had fused with fibroblasts to give unpigmented hybrids were themselves actually pigmented.

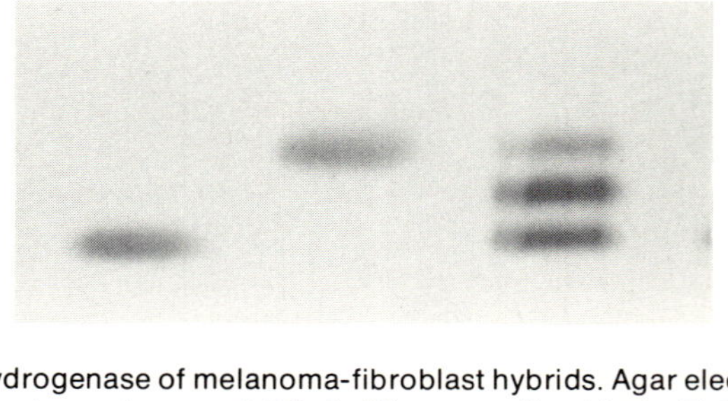

FIG. 2. Malate dehydrogenase of melanoma-fibroblast hybrids. Agar electrophoresis. From left: (1) Syrian hamster melanoma 3460–3; (2) mouse fibroblast; (3) 3460–3 x fibroblast hybrid.

The most difficult problem in studying the regulation of a function that is not expressed in the hybrids is to show that the genes for that function have been retained in the hybrids. Obviously the only definitive proof that the relevant genes were retained is the reappearance of the function. Experiments in which functions have returned will be discussed below. However, in the absence of this type of evidence, suggestive evidence can be obtained from karyotypic analysis. The melanoma cells had an average chromosome number of 51. One of the fibroblasts with which the melanoma cells were fused had an average chromosome number of 52. The hybrids had an average of 101 chromosomes, closely approximating the sum of the parental karyotypes (13). This made it highly unlikely that all of the hybrids were unpigmented as a result of the loss of the hamster chromosomes carrying the genes for pigment synthesis.

A general problem in studies of differentiation in hybrid cells involves the effect of the experiment itself on the expression of the differentiated function. Stated specifically, there could be something about (a) the fusion process itself, (b) the combination of the genomes of cells of different species, or (c) the increase in cell size and chromosome number, which prevents differentiated functions from being expressed in the hybrids. As will be described below, pigmented hybrids between melanoma cells and fibroblasts can be produced by altering the genome ratios (8, 15). This result rules out at once all three of these possibilities.

The conclusion from the initial studies was that the absence of pigment in the hybrids was not an artifact but was a specific effect possibly due to the action of a regulatory system for the control of pigment synthesis. It was suggested that the genome of a fibroblast produces a diffusible regulator substance that blocks the expression of pigment forming genes, possibly by preventing the synthesis of the enzyme dopa oxidase. Other mechanisms could certainly explain the observed results, but this one is perhaps the simplest and most straightforward.

Based on the results of experiments with glial cells, which suggested that the expression of differentiated functions in hybrid cells could be affected by the ratio of differentiated cell:undifferentiated cell genomes (see below), hybrids that contained two pigment cell genomes and one fibroblast genome were isolated (8). These hybrids were produced by fusing near tetraploid Syrian hamster melanoma cells ($3460\text{–}3^2$, derived from the same melanoma cells used in the preceding experiments) with mouse fibroblasts. These hybrids will be called P/P/F hybrids to indicate that they contain two pigment cell genomes and one fibroblast genome, in contrast to the previously isolated hybrids that contained only one genome of each type and will be referred to as P/F hybrids. The results of this cross were strikingly different from those of the earlier experiments. Approximately half of the P/P/F hybrid lines isolated were unpigmented, whereas the other half were darkly pigmented (see Fig. 1). Dopa oxidase assays were performed and the hybrids that were unpigmented had no detectable enzyme activity, whereas the pigmented hybrids had high activity, in some cases even higher than the parental melanoma cells.

Experiments were undertaken to determine why some, but not all, of the P/P/F hybrids were pigmented. Karyotypic analysis revealed little difference between the two types of P/P/F hybrids. Furthermore, a group of 10 enzymes present in both parental cells before fusion and not related to the differentiation of pigment cells was examined in the hybrids, and for all of the enzymes, the forms of both parents were present in the two types of P/P/F hybrids. (The enzymes were kindly analyzed by Dr. Frank Ruddle.) These results suggested that there were no great differences in the genetic complements of the two types of P/P/F hybrids.

Both unpigmented and pigmented P/P/F hybrids were subcloned in order to see whether one type of hybrid could segregate cells of the other type. It was found that unpigmented hybrids yielded only unpigmented subclones. On the other hand, pigmented hybrids produced both pigmented and unpigmented subclones. (By a mechanism that is not understood, the frequency with which a given pigmented hybrid segregated unpigmented subclones seemed to be genetically determined and was passed on to its pigmented progeny.) Karyotypic analyses of the pigmented and unpigmented subclones revealed little difference between them.

The isolation of pigmented hybrids in the gene dosage experiments indicated that the number of pigment cell genomes in the hybrids could affect the expression of the differentiated function. This suggested some quantitative aspect to the regulation of differentiation. However, the fact that only some of the P/P/F hybrids were pigmented made the results difficult to interpret. It could be hypothesized that all of the hybrids were initially unpigmented, and that the loss of some of the chromosomes of the undifferentiated parent resulted in some of the hybrids becoming repigmented. This hypothesis is consistent with neither the karyotypic similarity of the pigmented and unpigmented hybrids nor the observed segregation of unpigmented hybrids from pigmented ones.

Alternatively it could be hypothesized that all of the hybrids were initially pigmented and that during growth some of the hybrids became unpigmented. This is at least consistent with the observed pattern of segregation. In this case, the presence of pigment in the hybrids could be explained on the basis of a gene dosage effect. If the absence of pigment in the previously isolated P/F hybrids was due to the production by the fibroblast genome of a regulator substance that prevented pigmentation, then the appearance of pigment in the P/P/F hybrids could indicate that the fibroblast genome produced an amount of that regulator substance sufficient to control its own activity plus the activity of one, but not two, additional pigment cell genomes. If the appearance of pigment in the P/P/F hybrids is due to such an effect, the observed disappearance of pigmentation upon subcloning could be explained by the loss from the hybrids of a single chromosome that carried pigment synthesizing genes, thereby tipping the balance back in favor of the regulator substance. (Using a gene dosage hypothesis to explain the difference between P/F and P/P/F hybrids, one could just as easily hypothesize that the P/F hybrids were unpigmented because they lacked a sufficient concentration of some substance produced by the melanoma genome and necessary for pigmentation and that this substance attained the necessary concentration in the P/P/F hybrids.) Experiments are presently in progress to determine whether the dopa oxidase in the pigmented P/P/F hybrids is only of the hamster (melanoma) type, or whether there is also present the mouse type, which would be indicative of the activation of the fibroblast genome in the hybrids.

In the initial report on the absence of pigment synthesis in hybrids between pigment cells and fibroblasts, a number of questions were raised (12), and much of the subsequent work in the area has been along the lines indicated by those questions. Among the questions raised were the following: Are other differentiated functions suppressed in hybrids between differentiated and undifferentiated cells? Are the multiple functions characteristic of a given cell type coordinately controlled? Does the suppression of a differentiated function in the hybrids require the continued presence of certain chromosomes of the undifferentiated parent?

The review of the results of other systems presented above indicated that many, but not all, other differentiated functions are absent in hybrids between differentiated and undifferentiated cells. Studies with glial cells, kidney cells, and liver cells have provided evidence concerning the other two questions.

Regulation of Glial Cell Functions

Rat glioma cells, exhibiting a number of differentiated functions characteristic of glial cells—synthesis of the protein S100, high levels of the enzyme glycerol phosphate dehydrogenase (GPDH), and the inducibility of GPDH by hydrocortisone—were used in hybridization experiments (1, 10). Since the glial cells were characterized by several functions, it was possible to study the coordination of multiple functions in the hybrids. Near diploid (1s) rat glial cells (RG6A) were first fused with themselves, and near tetraploid (2s) "hybrids" ($RG6A^2$) were isolated. The 2s "hybrids" still retained the differentiated characteristics of the 1s parental glial cells. Both the 1s and 2s glial cells were hybridized with mouse fibroblasts. The hybrids containing one glial cell and one fibroblast genome will be referred to as G/F hybrids, and the hybrids containing two glial cell and one fibroblast genome will be referred to as G/G/F hybrids.

Immunological tests for the presence of S100 suggested that there had been at least a 90 to 95% reduction in S100 activity in the hybrids relative to the parental glial cells (1) (see Table 2). There was no significant difference between the G/F and G/G/F hybrids. The interpretation of the results was complicated by the fact that the material that reacted with the S100 antibody in the hybrids did not seem to be identical (in terms of complement fixation) to the S100 in the glial cells. At present, it cannot be said whether the low level of S100 activity in the hybrids was due to the presence of a modified S100 or to some cross-reacting (non-S100) material. It will be necessary to isolate the material from the hybrids in order to determine whether the synthesis of S100 had been suppressed in the hybrids or whether S100 was being produced but in a modified form.

TABLE 2. *Differentiated functions in glial cell–fibroblast hybrids*

Cell line	S100[a]	GPDH Baseline[a]	GPDH Induced
Fibroblast	5	7	7
Fibroblast x 1s glial cell hybrid (G/F)	8	23	25
Fibroblast x 2s glial cell hybrid (G/G/F)	9	87	110
1s glial cell	100	100	256
2s glial cell	65	201	359

[a] The levels of S100 and noninduced GPDH activity in the 1s glial cells are arbitrarily assigned the value of 100. Values are in terms of activity/mg protein.

The hybrids were also tested for GPDH baseline (noninduced) activity, and the inducibility of GPDH by hydrocortisone (10). As shown in Table 2, the baseline level of GPDH in the G/F hybrids was reduced approximately 75 to 80%. When the cells were treated with hydrocortisone, there was no increase in GPDH activity in the G/F hybrids, whereas the glial cells showed a 150% increase in GPDH activity. This suggested that a step in the hormone mediated enzyme induction was being blocked in the hybrids, presumably by some factor produced by the fibroblast genome. Experiments are in progress to further study the mechanism of control of inducibility. One possibility that must be considered is that the absence of inducibility in the hybrids is correlated with the absence of a hydrocortisone receptor and that the observed absence of GPDH inducibility in the hybrids represents a change in hormone responsiveness rather than a specific effect on the inducibility of this enzyme per se.

In contrast to the G/F hybrids, the G/G/F hybrids had a baseline GPDH activity almost as high as that of the 1s glial cells. However, unlike the glial cells, there was only a small (approximately 25%) increase in GPDH activity when the G/G/F hybrids were exposed to hydrocortisone. The baseline GPDH activities in the G/G/F hybrids and 1s glial cells were comparable, whereas the induced levels were markedly different, suggesting that baseline activity and inducibility are controlled independently. On the other hand, the fact that there was some increase in GPDH activity in the G/G/F hybrids following hydrocortisone treatment, whereas there was no increase in the G/F hybrids, suggested a gene dosage effect as was described above for pigment synthesis. It is possible that the presence of two glial cell genomes in the G/G/F hybrids had overcome the barrier to inducibility caused by the fibroblast genome, whereas the single glial cell genome in the G/F hybrids was blocked in its inducibility by the fibroblast genome.

Information on the coordination of multiple functions was obtained in this series of experiments by comparing S100 and GPDH levels in hybrids with different genome ratios (see Table 2). The addition of either one or two glial cell genomes to a fibroblast genome had no effect on the S100 level. (As previously indicated, the significance of the low level of S100 activity in the hybrids is not known.) In contrast, the addition of one glial cell genome to a fibroblast genome (in G/F hybrids) caused a threefold increase in GPDH (baseline) activity, and the addition of a second glial cell genome (in G/G/F hybrids) resulted in a further fourfold increase. These results did not indicate the specific mechanisms involved in the control of S100 and GPDH activities, but the results did suggest that the two proteins characteristic of glial cells are controlled independently in hybrids between glial cells and fibroblasts. The conclusions from these studies on the control of enzyme inducibility and the lack of coordination of multiple functions have been confirmed in studies on hybrids with liver cells (see below).

Regulation of a Kidney Specific Function

It was suggested above that the absence of a differentiated function in hybrids between differentiated and undifferentiated cells could be due to the activity of a genetic element that suppressed the expression of the function. Very strong support for this hypothesis would be obtained from the reappearance of the differentiated function in the hybrids correlated with the loss of a specific chromosome of the undifferentiated parent. (It should be pointed out that, depending on the mechanism of regulation, the differentiated function would not necessarily reappear following the loss of a genetic element that caused its suppression. Thus the reappearance of the differentiated function would support the hypothesis of a genetic regulatory element, but the failure of reappearance would not contradict the hypothesis.)

In one set of experiments, a correlation was reported between the loss of a specific chromosome and the reappearance of a differentiated function. The experiments will be described in detail, even though the results could not be confirmed, because they serve to indicate the types of information that could be obtained. In these experiments, mouse renal adenocarcinoma cells (RAG), which synthesized the kidney specific esterase ES-2, were hybridized with mouse and human fibroblasts (18). ES-2 activity could not be detected in the hybrids with mouse fibroblasts. In contrast seven out of eight hybrids with the human fibroblasts did produce ES-2. The one hybrid with the human fibroblasts that did not produce ES-2 was subcloned, and one subclone that exhibited ES-2 activity was isolated. Karyological analyses suggested that the subclone that had regained ES-2 activity had lost a specific human chromosome, number 10, whereas those hybrids that did not exhibit ES-2 activity retained the human chromosome 10. (Recent experiments with hybrids between RAG cells and human lymphocytes did not confirm the correlation between the human chromosome 10 and ES-2 activity. Similar studies indicating a correlation between tyrosine aminotransferase inducibility and the human X chromosome are presented by Croce and Koprowski in this volume.)

Results of this nature are important for a number of reasons. The reappearance of a differentiated function after the loss of some of the chromosomes of the undifferentiated parent demonstrates that the genes for the differentiated function were present in the hybrids at the time when the activity was not observed. The correlation between the reappearance of activity and the loss of a specific human chromosome further suggests that a gene on that chromosome produces a diffusible regulator substance that blocks the expression of the differentiated function. Furthermore, the results suggest that this regulator substance is produced continually, rather than being produced only once and having a persistent effect, since the suppression of the function is apparently reversed when the specific chromosome is lost. Finally, it may be suggested that this regulator substance affects only the expression of the differentiated function and not the potential of a cell to carry out the function, since the hybrids maintain throughout many generations the ability to carry out the function in the absence of its expression and begin expressing the function again following chromosome segregation. This distinction between potential and expression may be relevant to the embryological distinction between determination and differentiation.

Regulation of Liver Cell Functions

In general, studies on the reappearance of functions in hybrid cells can best be performed with hybrids between human cells and rodent cells, since the human chromosomes are preferentially lost from the hybrids (32). However, in a series of rodent-rodent hybrids of which one of the parents was a rat hepatoma cell expressing several differentiated functions, an unusual chromosome segregation occurred, which allowed the study of the reappearance of multiple functions. In these experiments rat hepatoma cells (FU 5–5) that expressed a number of liver specific functions, including the synthesis of aldolase B and liver alcohol dehydrogenase (ADH), as well as a high-baseline activity of tyrosine aminotransferase (TAT) and its inducibility by hydrocortisone, were hybridized with buffalo rat liver cells, which, in spite of their derivation from liver tissue, expressed none of these tissue specific functions. All of these liver specific functions were absent in the hybrids (3, 4, 27). However, upon chromosome loss, various functions were regained by some of the hybrids. It was observed that the hybrids could regain one function without regaining the others; e.g., one hybrid was characterized by the reappearance of TAT inducibility (31) without the return of a high-baseline TAT activity and also without the return of ADH activity, while another hybrid regained ADH activity (3) without the return of high-baseline TAT activity or TAT inducibility. These results suggested that TAT baseline and inducibility are independently controlled and also that the different liver specific functions are not coordinately controlled in the hybrids (*cf.* the conclusions of the experiments with glial cells presented above). Since both parental cells were derived from rats, it was not possible to make any correlation between the return of differentiated functions and the loss of specific chromosomes.

The rat hepatoma cells were also characterized by the production of albumin. These cells were hybridized with mouse fibroblasts in gene dosage experiments similar to those described above for pigment and glial cells. Hybrids containing either one liver cell and one fibroblast genome (L/F hybrids) or two liver cell and one fibroblast genome (L/L/F hybrids) were isolated. In the L/F hybrids, a small amount of albumin was produced, and immunological tests indicated that the albumin was of the rat (hepatoma) type. Albumin production was also observed in the L/L/F hybrids. However, in some of the L/L/F hybrids, albumin of the mouse (fibroblast) type was also detected (25). The presence of mouse albumin in some L/L/F hybrids but not in L/F hybrids could be explained on the basis of a gene-dosage effect as discussed above for pigment synthesis. The results clearly indicated that the genes for albumin synthesis, which, as a result of differentiation, were not being expressed in fibroblasts, were retained by the fibroblasts in such a state that they could be reactivated. The results are consistent with the hypothesis that the genome of a liver cell produces a diffusible regulator substance that functions to initiate the expression of the genes for albumin synthesis. As with the other experiments discussed above, alternative explanations could also account for the results.

DISCUSSION

The results of the experiments on the regulation of differentiated functions in hybrid cells do not provide any unequivocal demonstration of the mechanisms of gene regulation in mammalian cells. However, a framework can be set up with which the results are consistent. The results of the experiments suggest that mammalian cells use diffusible regulator substances to control the expression of many of their differentiated functions. The results are consistent with the existence of at least two different types of regulator substances, one type that prevents the expression of differentiated functions (as suggested by the experiments on melanin synthesis) and another type that initiates the expression of differentiated functions (as suggested by the experiments on albumin synthesis). It is clear, however, that both the suppression and induction of differentiated functions in hybrid cells can be explained on the basis of quantitative variations of only one type of mechanism.

The results of the gene dosage experiments suggest that the regulator substances are produced in rather limited quantities, since changing the genome ratios in hybrids by a factor of two can alter the expression of differentiated functions. The results of the experiments on the reappearance of functions following chromosome segregation suggest that at least some of the regulator substances are being continually produced and that their effect is limited to the time when they are present. The results also suggest that these regulator substances may be involved in the control of the immediate expression of differentiated functions and not in the control of the potential of a cell to carry out a given function. Finally, the results of the experiments with cells characterized by more than one function suggest that the multiple functions expressed by a given cell type are not coordinately controlled, at least not in terms of the temporal expression of the functions. The results obviously do not rule out coordination at the level of the determination of potential.

The above conclusion must be regarded as speculative and tentative. The ultimate proof of the mechanisms of regulation in mammalian cells may depend upon the refinement of techniques of biochemical analysis, so that single genes and regulator substances can be purified and their interactions studied. However, the genetic experiments of the type described in this article can indicate the types of interactions that are probably occurring. The theoretical framework set up above may be of value not because it is necessarily correct but because it can suggest the direction of further experimentation.

REFERENCES

1. Benda, P., and Davidson, R. L. (1971): Regulation of specific functions of glial cells in somatic hybrids. I. Control of S100 protein. *J. Cell. Physiol. 78*:209–216.
2. Benedict, W., Nebert, D., and Thompson, E. B. (1972): Expression of aryl hydrocarbon hydroxylase induction and suppression of tyrosine aminotransferase induction in somatic cell hybrids. *Proc. Nat. Acad. Sci. 69*:2179–2183.
3. Bertolotti, R., and Weiss, M. (1972): Expression of differentiated functions in hepatoma cell hybrids. VI. Extinction and re-expression of liver alcohol dehydrogenase. *Biochimie 54*:195–201.
4. Bertolotti, R., and Weiss, M. *(in press):* Expression of differentiated functions in hepatoma cell hybrids. V. Re-expression of aldolase B following chromosome loss. *J. Cell. Physiol.*
5. Coffino, P., Knowles, B., Nathenson, S., and Scharff, M. (1971): Suppression of immunoglobulin synthesis by cellular hybridization. *Nature 231*:87–90.

6. Colten, H., and Parkman, R. (1972): Biosynthesis of C4 (fourth component of complement) by hybrids of C4-deficient guinea pig cells and HeLa cells. *Science 176*:1029–1031.
7. Davidson, R. L. (1969): Regulation of melanin synthesis in mammalian cells, as studied by somatic hybridization. III. A method of increasing the frequency of cell fusion. *Exp. Cell Res. 55*:424–426.
8. Davidson, R. L. (1972): Regulation of melanin synthesis in mammalian cells: Effect of gene dosage on the expression of differentiation. *Proc. Nat. Acad. Sci. 69*:951–955.
9. Davidson, R. L. (1973): *Somatic Cell Hybridization: Studies on Genetics and Development*, Addison-Wesley, Reading, Mass.
10. Davidson, R., and Benda, P. (1970): Regulation of specific functions of glial cells in somatic hybrids. II. Control of inducibility of glycerol phosphate dehydrogenase. *Proc. Nat. Acad. Sci. 67*:1870–1877.
11. Davidson, R. L., and Yamamoto, K. (1968): Regulation of melanin synthesis in mammalian cells, as studied by somatic hybridization. II. The level of regulation of 3,4-dihydroxyphenylalanine oxidase. *Proc. Nat. Acad. Sci. 60*:894–901.
12. Davidson, R. L., Ephrussi, B., and Yamamoto, K. (1966): Regulation of pigment synthesis in mammalian cells, as studied by somatic hybridization. *Proc. Nat. Acad. Sci. 56*:1437–1440.
13. Davidson, R. L., Ephrussi, B., and Yamamoto, K. (1968): Regulation of melanin synthesis in mammalian cells, as studied by somatic hybridization. I. Evidence for negative control. *J. Cell. Physiol. 72*:115–128.
14. Finch, B., and Ephrussi, B. (1967): Retention of multiple developmental potentialities by cells of a mouse testicular teratocarcinoma during prolonged culture *in vitro* and their extinction upon hybridization with cells of permanent lines. *Proc. Nat. Acad. Sci. 57*:615–621.
15. Fougere, C., Ruiz, F., and Ephrussi, B. (1972): Gene dosage dependence of pigment synthesis in melanoma-fibroblast hybrids. *Proc. Nat. Acad. Sci. 69*:330–334.
16. Jami, J., Failly, C., and Ritz, E. (1973): Lack of expression of differentiation in mouse teratoma-fibroblast somatic cell hybrids. *Exp. Cell Res. 76*:191–199.
17. Kao, F., and Puck, T. (1972): Genetics of somatic mammalian cells: Demonstration of a human esterase activator gene linked to the ade B gene. *Proc. Nat. Acad. Sci. 69*:3273–3277.
18. Klebe, R., Chen, T., and Ruddle, F. (1970): Mapping of a human regulator element by somatic cell genetic analysis. *Proc. Nat. Acad. Sci. 66*:1220–1227.
19. McMorris, F. A., Kolber, A., Moore, B., and Perumal, S. *(in press):* Differentiated functions in neuroblastoma hybrids. III. Expression of the neuron-specific protein, 14-3-2, and steroid sulfatase.
20. Minna, J., Nelson, P., Peacock, J., Glazer, D., and Nirenberg, M. (1971): Genes for neuronal properties expressed in neuroblastoma–L cell hybrids. *Proc. Nat. Acad. Sci. 68*:234–239.
21. Minna, J., Glazer, D., and Nirenberg, M. (1972): Genetic dissection of neural properties using somatic cell hybrids. *Nature 235*:225–231.
22. Mohit, B. (1971): Immunoglobulin G and free kappa chain synthesis in different clones of a hybrid cell line. *Proc. Nat. Acad. Sci. 68*:3045–3048.
23. Mohit, B., and Fan, K. (1971): Hybrid cell line from a cloned immunoglobulin producing mouse myeloma and a nonproducing mouse lymphoma. *Science 171*:75–77.
24. Periman, P. (1970): IgG Synthesis in hybrid cells from an antibody-producing mouse myeloma and an L cell substrain. *Nature 228*:1086–1087.
25. Peterson, J., and Weiss, M. (1972): Expression of differentiated functions in hepatoma cell hybrids: Induction of mouse albumin production in rat hepatoma-mouse fibroblast hybrids. *Proc. Nat. Acad. Sci. 69*:571–575.
26. Rintoul, D., Colofiore, J., and Morrow, J. (1973): Expression of differentiated properties in fetal liver cells and their somatic cell hybrids. *Exp. Cell Res. 78*:414–422.
27. Schneider, J., and Weiss, M. (1971): Expression of differentiated functions in hepatoma cell hybrids. I. Tyrosine aminotransferase in hepatoma-fibroblast hybrids. *Proc. Nat. Acad. Sci. 68*:127–131.
28. Silagi, S. (1967): Hybridization of a malignant melanoma cell line with L cells *in vitro*. *Cancer Res. 27*:1953–1960.
29. Sonnenschein, C., Richardson, U., and Tashjian, A. (1971): Loss of growth hormone pro-

duction following hybridization of a functional rat pituitary cell strain with a mouse fibroblast line. *Exp. Cell Res.* *69*:336–344.
30. Sparkes, R., and Weiss, M. (1973): Expression of differentiated functions in hepatoma cell hybrids: Alanine amino transferase. *Proc. Nat. Acad. Sci.* *70*:377–381.
31. Weiss, M., and Chaplain, M. (1971): Expression of differentiated functions in hepatoma cell hybrids: Reappearance of tyrosine aminotransferase inducibility after the loss of chromosomes. *Proc. Nat. Acad. Sci.* *68*:3026–3030.
32. Weiss, M., and Green, H. (1967): Human-mouse hybrid cell lines containing partial complements of human chromosomes and functioning human genes. *Proc. Nat. Acad. Sci.* *58*: 1104–1111.
33. Wiebel, F., Gelboin, H., and Coon, H. (1972): Regulation of aryl hydrocarbon hydroxylase in intraspecific hybrids of human, mouse, and hamster cells. *Proc. Nat. Acad. Sci.* *69*: 3580–3584.

DISCUSSION

Thompson: Why do you feel that these experiments represent evidence for repressors of differentiated functions rather than for any one of a multitude of other mechanisms, which one could readily imagine, in which nonspecific effects would cause failure of expression of a given phenotype in hybrids?

Davidson: I would rather not use the term "repressor." However, I think that the results are suggestive of specific control mechanisms. The more functions that are studied, the better the chance that the results are not due to nonspecific effects. Many differentiated functions have now been observed to disappear in hybrids. It would be rather unusual, I think, if there were some nonspecific mechanism that affected all of these different functions in the same way. The strongest evidence for the existence of regulatory mechanisms that specifically prevent the expression of differentiated functions would come from studies of the reappearance of a function, especially where the reappearance of the function can be correlated with the loss of a specific chromosome. Two such cases have now been reported.

Siniscalco: Have you ever tried to synchronize the cells that show differences in the production of melanin? Could the differences be related to the fact that, during the growth cycle of the melanoma cells, the dopa oxidase activity varies greatly?

Davidson: If I understand what you are suggesting, there is no reason why the regulation could not involve the cyclical variations that we observed. It is possible that the "suppressed" parent is always maintained at the low level whereas the differentiated parent is allowed to go through a cycle where it does show the differentiated function.

There is one point that may not be clear. The variations we described were not cyclic variations during the cell cycle, but cyclic variations in a nonsynchronized population over the entire growth cycle from inoculation to confluence.

Siniscalco: I am suggesting that what you call suppression may not really be so. If you could look at the cells in a synchronized culture at a certain moment of the cell cycle, they might all produce melanin.

Davidson: This does not explain why enzyme activity is not seen in nonsynchronized cultures. However, the role of cyclical type variations can not be ruled out.

Attardi: Why don't you consider the possibility of membrane effects in the hybrid cells? It appears that there is rapid redistribution of membrane components in fused cells. A new membrane is formed in the hybrids and some of the effects on the expression of functions in hybrids might be the result of changes in the membrane.

Davidson: We can not rule out the possibility of membrane changes. However, the dosage experiments show quite clearly that simply the mixture of the membranes of two different cells is not a sufficient explanation for the absence of differentiated functions in hybrid cells.

The mechanism I proposed is what I consider to be the easiest way of explaining the results. There are obviously many other possibilities, and many more experiments are needed.

Shows: Could one mechanism involve the alteration of one of the gene products? For example, acid phosphatase possesses sialic acid moieties and after starch gel electrophoresis there are two major zones of acid phosphatase activity. In some man-mouse hybrids, one of the mouse zones disappears, while the other zone increases in activity. When neuraminidase is added to hybrid homogenates that have lost one of the mouse acid phosphatases, the zone that remains is converted to the zone that was lost. These results suggest that a human sialyl transferase or other processing enzyme may be present in the hybrids and may convert one form of mouse acid phosphatase into the other, thereby causing an apparent extinction of one form of the mouse enzyme (Shows, T., and Lalley, P., *in press*).

Davidson: We considered molecular alteration that would lead to decreased activity in the case of S100, where the hybrids revealed an activity that did not react with anti-S100 like the native protein. I would not expect this to be a common method of controlling enzyme activity, i.e., altering proteins after their synthesis. However, there is no evidence to demonstrate that it is not the case.

Ephrussi: The results of the experiments with glial cells performed by Davidson and Benda in my laboratory raised in our minds the possibility that extinction may not result from a block in transcription, but rather from the modification of a synthesized molecule. I think this is a very real possibility. In other words, whether the phenomenon that I call "extinction" and Davidson calls "suppression" is due to regulation at the level of transcription or at some other level is completely unknown. The major difficulty, especially in the case of melanoma cells in which the enzyme characteristic of the differentiated function is poorly characterized, is the detection of an inactive enzyme.

I would also like to comment on Attardi's remark. I take the question of membranes and the cell surface very seriously, particularly in the case of melanoma cells. I think there is a very definite correlation between the ability of a cell to produce dopa oxidase and its surface properties. This is suggested by the fact that pigmentation, like many other differentiated functions, is suppressed in the presence of BUDR. The first effect of BUDR, even before the disappearance of pigment, is a change in cell morphology. Under the influence of BUDR, the cells become definitely fibroblastic, and they flatten out. They increase in volume. Their properties of adherence to the substrate are very different. Furthermore, with the hybrids with melanoma cells, the differences in pigmentation among the hybrids are reflected in the surface properties of the cells, as diagnosed simply by their morphology and affinity for the substrate.

Davidson: I think one probably could ascribe the results to any particular level one wanted to. At the present time, we really have no proof at all of the level at which regulation is occurring. I think the results do suggest, however, that regulation is occurring at some level. I inherently distrust any morphological changes in cells, because we have seen variations that are not correlated with changes in function.

Ephrussi: I agree with most of what Davidson said, but I think that this question of membranes remains a fundamental one. Morphology, of course, is not a reliable criterion by itself. However, when it is systematically correlated with the change of another measurable function, it is much more impressive.

I would like to quote some unpublished work from my lab on one of the characteristics of the Syrian hamster melanoma mentioned by Davidson, namely, you don't find amelanotic cells in the population, unlike other melanomas. The quoted frequency of amelanotic cells was less than one per 30,000. We were very puzzled by this, and we tested the effect of substrate attachment on pigmentation. We plated the melanoma cells out in bacteriological Petri dishes, in which the cells usually don't attach, and made two surprising observations. The first was that, in the presence of BUDR, all cells that we have tried will attach to bacteriological Petri dishes.

Second, from the rare melanoma cells that do attach in these Petri dishes without BUDR, we easily recovered amelanotic mutants. These cells were again characterized by cell surface changes, in addition to the stable disappearance of melanogenesis.

Davidson: I would like to just mention one experiment that I think indicates some of the problems in working with BUDR and cell morphology. We have isolated a mutant melanoma cell that requires BUDR. Normally, if a melanoma cell is exposed to BUDR, the cells spread out and become much more fibroblastic. However, the BUDR-dependent cells in the presence of BUDR look like the normal cells in the absence of BUDR. When the BUDR is taken away from the dependent cells, they become more spread out and resemble normal cells in the presence of BUDR.

Harris: There is an old paper in the literature correlating pigmentation in melanoma cells in culture with growth rates. I wonder whether in those cases where you see both pigmented and unpigmented hybrids, there is such a correlation? Silagi also noted that pigmentation was restored to amelanotic cells after treatment with cytosine arabinoside, which could conceivably select for variants with slower growth rate. These could be completely independent phenomena, but have you any data that would enable us to see whether there is a correlation?

Davidson: We have tested cytosine arabinoside and observed no effect on pigmentation in hybrids. As far as growth rates are concerned, we have not observed any significant difference between the generation times of the pigmented and unpigmented hybrids.

Pontecorvo: I should like to put in a plea that you standardize your terms. I have been brought up reading papers about luxury and household functions and that was perfectly clear to me; then, extinction, suppression, re-expression. Decide among yourselves and please use one set of terms.

Ephrussi: I am prepared to stick to the terms I had previously proposed. I would like to make just one point about the choice of the term extinction. Extinction is formally or operationally equivalent to recessiveness. However, in differentiation, we don't know what we are dealing with. We all believe that real genetic changes, like mutations, are not involved. Therefore, I proposed to use the term extinction rather than to speak of recessiveness.

Littlefield: In the list of functions expressed in hybrids, why is collagen not included?

Davidson: The functions I included in the table were for those cases in which a hybrid was isolated between a cell that expressed a function and a cell that did not. In the case of collagen, both parental cells synthesized collagen, but at different rates. I think we may possibly get into an entirely different type of regulation when we look at this.

Silagi: I would like to clarify one point lest anyone be confused. My experiments with cytosine arabinoside (ara C) were done with a melanoma line, which was not producing any melanin at all. It had nothing to do with hybrids [Silagi, S. (1969): *J. Cell Biol. 43*:263].

Subsequently, when I derived unpigmented hybrids between mouse melanoma cells and A9 cells, I also again tested them in the same fashion [Silagi, S. (1967): *Can. Res. 27*:1953]. We tried to force them to make pigment by growing them on agar or treating them with ara C. Since some of the hybrids were tumorogenic, we also checked to see whether they would form pigmented tumors [Ruddle, F., Chen, T., Shows, T., and Silagi, S. (1970): *Exp. Cell Res. 60*:139]. These hybrids did not express pigment in any of these cases, not with ara C, not with agar, and not in the tumors that were formed.

Minna: I would like to come back to Siniscalco's point, since it can be demonstrated that clonal lines of cells show regulation of differentiated functions within a clone (intra- not interclonal variation). It seems entirely possible that the results observed with the hybrids, in terms of what functions are expressed, could very well be due to the state of differentiation of the parental cell at the time of fusion. It is quite possible that when you introduce 2S cells into the cross, either in the case of 2S melanoma cells or the highly aneuploid neuroblastoma cells that I use, you can catch cells in a state of differentiation, freeze that state by fusion and then have it expressed in the hybrid cells.

The state of the differentiated parent at the time of fusion may thus be one of the determining factors in whether or not the hybrids can express the function.

Davidson: You would then expect that sometimes the cells would fuse while the differentiated parent was in the right phase and sometimes in the wrong phase. However, we have examined hundreds of hybrids with 1S pigment cells and have never seen any pigmented hybrids. Furthermore, the type of clonal regulation you are referring to occurs with neuroblastoma cells and with TAT inducible hepatoma cells. We have looked at over 30,000 clones of the melanoma cells and we do not see this type of intraclonal variation.

Coon: If Minna is correct, this makes the specific prediction that if you were to hybridize BUDR-treated pigment cells, the hybrids would not be pigmented no matter what their ploidy.

Davidson: Such an experiment has not been performed.

Somatic Cell Hybridization, edited by
Richard L. Davidson and Felix de la Cruz.
Raven Press, New York © 1974.

Reexpression of Liver Specific Enzymes in Hepatoma Cell Hybrids

Mary C. Weiss

Centre de Génétique Moléculaire, Gif sur/Yvette, France

My co-workers (Roger Bertolotti, Michèle Chaplain, Robert Sparkes) and I have been studying the properties of hybrids with hepatoma cells. The rat hepatoma cells (FU5–5), which we have been studying, are well differentiated, express many liver specific functions, and grow and clone well. With phase-contrast optics, the hepatoma cells have a dark cytoplasm, a pale nucleus, and frequently one large centrally placed nucleolus (which is characteristic of liver cells).

The liver enzymes that we have analyzed include a high baseline of tyrosine aminotransferase (TAT) and its inducibility with corticosteroids, a high baseline of alanine aminotransferase (AAT) and its inducibility, liver alcohol dehydrogenase (L-ADH) and aldolase B. I will discuss the patterns of reexpression of these liver specific enzymes in hybrid cells.

The rat hepatoma cells were hybridized with cells of a diploid epithelial line, BRL-1, derived from the liver of a young Buffalo rat by Coon. The exact origin of these cells remains unclear since they express none of the liver specific functions we have examined. (The BRL cells appear to be derived from relatively small cells present in liver cell suspensions that have a very clear cytoplasm and lack the large centrally placed nucleolus. Such cells are probably not differentiated hepatocytes but represent some type of liver epithelial cells.)

In morphology, hybrids between the hepatoma cells and BRL closely resemble BRL cells (see Fig. 1c). The hybrids contain the complete chromosomal complements of both parents and lack the four liver specific enzymes mentioned above.

From these hybrids, which like BRL-1 cells are highly contact inhibited, we have isolated a number of "variant" subclones, which grow in multiple layers like the hepatoma cells, and which have lost very large numbers of chromosomes. The initial hybrid cells contained 91 to 93 chromosomes while the "variant" subclones contain only 55 to 63 chromosomes. One of these hybrid subclones, BF5–1–1, reexpressed TAT inducibility (5), and had a moderate baseline activity of AAT (4). These cells have 60 to 66 chromosomes. A subclone derived from BF5–1–1, called BF5–1–1–α, showed

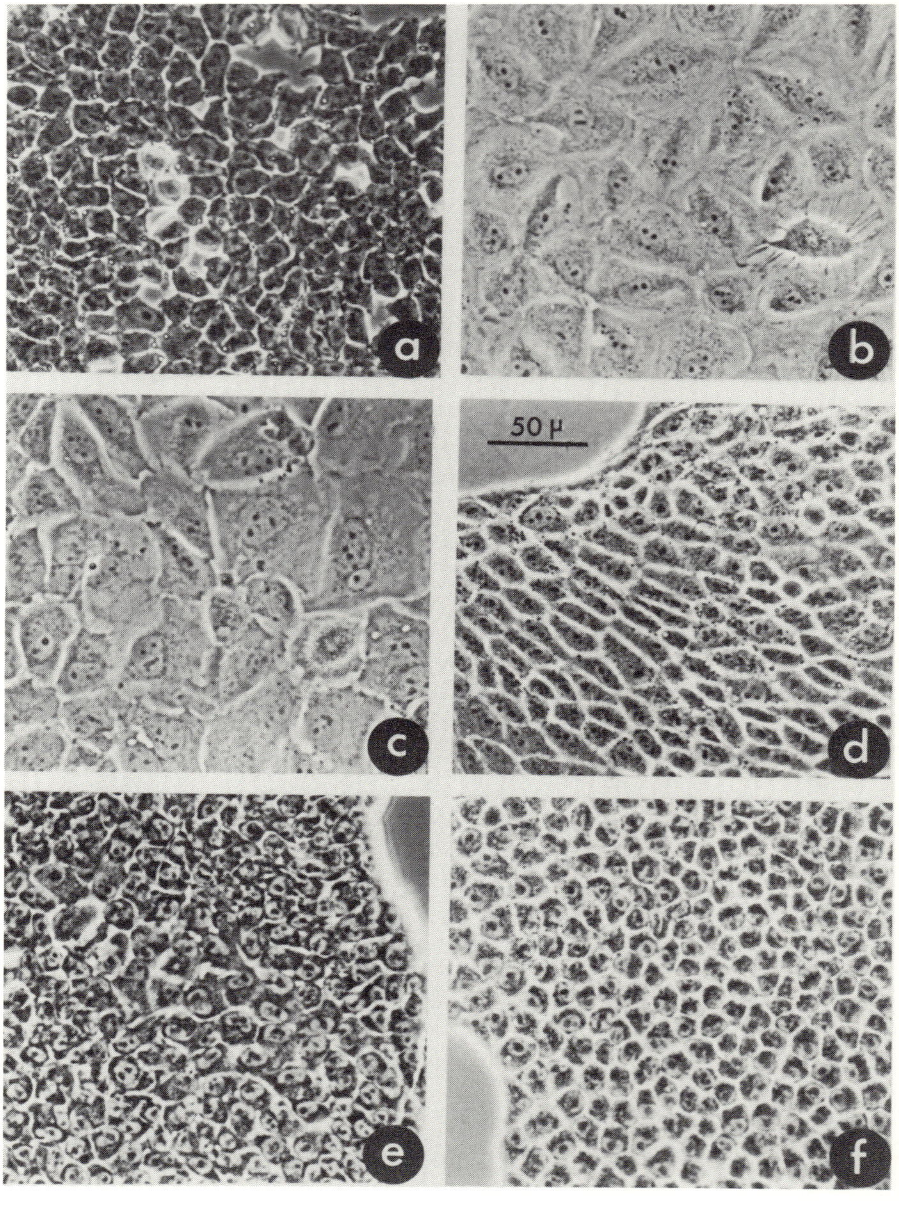

FIG. 1. Phase contrast photomicrographs of living cultures of: (a) Fu5–5, note the dark granular cytoplasm and the large centrally placed nucleoli; (b) BRL–1; (c) BF5, the original hybrid clone which contains the complete chromosome sets of both parents; (d) hybrid subclone BF5–1–1; (e) hybrid subclones BF5–γ–4; and (f) BF5–1–1–α, both of which show reexpression of morphological characteristics of Fu5–5 cells.

reexpression of high basal and induced levels of AAT (4), of aldolase B (1, 3) and a little liver-alcohol dehydrogenase. A third (independent) subclone BF5–γ–4 has also lost many chromosomes and is characterized by the absence of TAT baseline and inducibility and by the presence of aldolase B (1, 3), liver alcohol dehydrogenase (2), and high AAT baseline and inducibility (4) (see Fig. 1 and Table 1 for morphological and biochemical characteristics of the hybrids).

TABLE 1. *Chromosome numbers and liver specific enzymes in parental and hybrid cells*

Cell line	Mean no. of chromosomes	TAT[a]		AAT		L-ADH[b]	Ald. B.
		B[c]	I[c]	B[c]	I[c]		
Fu5–5	52(51–53)	++	++	++	++	++	++
BRL–1	42	—	—	—	—	—	—
BF5	92(91–93)	—	—	—	—	—	—
BF5–1–1	63(60–66)	—	+	+	—	—	—
BF5–1–1–α	60(57–62)	—	+	++	++	±	++
BF5–γ–4	55(54–55)	—	—	++	++	+	++

[a] TAT (tyrosine aminotransferase), AAT (alanine aminotransferase).
[b] L-ADH (liver form of alcohol dehydrogenase), Ald. B (aldolase B).
[c] B (baseline), I (induced: 10^{-6} M and 10^{-7} M dexamethasone for TAT and AAT, respectively).
+ Specific activity of reexpressed enzyme is less than that of Fu5–5.
++ Specific activity of reexpressed enzyme is equal to or greater than that of Fu5–5.

Although two of these subclones come close to reconstituting the original phenotype of the hepatoma parent cell, some differences are observed. For example, in the case of TAT reexpression, we have observed different induction kinetics, depending on the position of the cells in the growth cycle.

Bertolotti has done some additional experiments with aldolase B (1, 3). The initial hybrids, which were contact inhibited and contained all the chromosomes of both parents, were injected in large numbers into rats. Tumors arose that had lost many chromosomes and they nearly always reexpressed aldolase B. Moreover, aldolase B–producing tumors were formed by injection of cells of a subclone that reexpressed only TAT inducibility. One tumor formed by this hybrid subclone, when grown *in vitro,* was obviously heterogeneous, containing two very different types of cells: those with morphological characteristics of the cells originally injected (Fig. 1d), as well as a new cell type, similar in appearance to those shown in Fig. 1e. By picking and pooling numerous colonies of the two cell types, we were able to show a positive correlation between the reexpression of hepatoma-like morphology and aldolase B.

Similar patterns of independent reexpression of liver specific enzymes have been observed in hybrids between Chinese hamster cells and rat hepatoma cells (Bertolotti, Sparkes, and Weiss, *unpublished observations*). For example, a particular hybrid clone may reexpress one function, whereas subclones of that hybrid may reexpress additional functions. In all cases, however, slight differences in the expression of the liver enzymes between the hybrids and the parent hepatoma cells are found. For example, in some of the Chinese hamster rat hepatoma hybrids, both baseline and inducibility of TAT are reexpressed, but, when the kinetics of inducibility are studied, there is a dip in the curve, which is not found under similar conditions for the hepatoma cells.

These observations may have importance for understanding gene regulation. In hybrid cells that reexpress a function, combinations of genes exist that are not present in the parental cell. Certain of these gene combinations could modify or modulate the expression of the specific function. Hybrid cells may provide, therefore, model systems for the study of mechanisms that can modify the expression of differentiated functions.

I would like to discuss some of the interpretations that can be drawn from studies of the expression of differentiated functions by somatic hybrids. Extinction appears to distinguish between luxury (nonessential) and household (essential) functions. There are at present no examples of extinction of essential functions. Extinction has been shown to be subject to quantitative dosage effects, which is almost certainly biologically significant (see Davidson, *this volume*).

Reexpression of differentiated functions indicates that extinction is not due to loss of genetic material or to the fusion of undifferentiated variants that may exist in the initial population of the differentiated parental cells. Reexpression also shows the stability of the epigenotype of the differentiated parent. In the hybrid cells, the individual functions are not expressed for many generations. However, they can be reexpressed one at a time or several together. Finally, the phenomenon of reexpression shows that the maintenance of extinction requires the continuous presence of a certain chromosome or chromosomes of the nonexpressing parent.

What is the significance of the independence of reexpression? It demonstrates that, in the hybrid cells, there are multiple elements capable of regulating the expression of the various functions of a single tissue type. Note that this conclusion is not necessarily applicable to the hepatoma parent. The action of the individual elements responsible for extinction must be relatively specific since reexpression can be independent: if there were nonspecific blockage of a group of functions, reexpression would be coordinate.

What are the major questions in this area? We do not know whether reexpression requires the loss of more than one chromosome or chromosome pair of the nonexpressing parent. In the nonexpressing parent or hybrid cell, is there a hierarchy of regulatory elements, such that several chromosomes have to be lost before reexpression occurs? In those cases where reexpression occurs in the hepatoma hybrids, the reduction in chromosome number is very extensive. The hybrid cells that reexpress liver specific enzymes have only 5 to 10 more chromosomes than the original hepatoma parent cell.

Secondly, we do not know whether the elements of the nonexpressing parent cell which lead to extinction have anything to do with the regulation of that function in the nonexpressing parental cell. This question may be answered when, in cases of reexpression, it is possible to determine whether or not the homologous differentiated product of the nonexpressing parental cell is also expressed. The extensive chromosome loss in these hybrids may prohibit such experiments, however.

Finally, in all cases where reexpression has been observed, the hybrids were derived from the fusion of cells that differ in their histogenetic origins. One would like to know whether the patterns of extinction and reexpression differ depending upon the histogenetic origin of the parental cells.

I would like to discuss briefly some preliminary results that have a bearing on this question (Deschatrette, J., and Weiss, M., *unpublished experiments*). We have systematically attempted to isolate cells that differ from the differentiated hepatoma cells. A number of variant clones have been isolated, all characterized by the simultaneous disappearance of all four liver specific functions. One of these dedifferentiated hepatoma clones has been crossed with the original differentiated hepatoma clone. To our surprise, all eight hybrid clones studied showed extinction of all four enzymes. This is a hybrid between two hepatoma cells, one expressing and one not expressing the liver specific functions. Whatever mechanism caused the dedifferentiation of these hepatoma cells appears to be capable of extinguishing the differentiated functions of a cell of similar origin. Of the numerous subclones studied to date, two reexpress coordinately all four or three out of four of the liver specific functions, and two subclones reexpress only one function.

REFERENCES

1. Bertolotti, R., and Weiss, M. C. (1972): Aldolase in hepatoma cell hybrids: The extinction of the hepatic form and its re-expression following loss of chromosomes. In: *Cell Differentiation*, Proc. of the 1st Int. Conf. on Cell Differentiation (eds., R. Harris and D. Viza), p. 202–205, Munksgaard, Copenhagen.
2. Bertolotti, R., and Weiss, M. C. (1972): Expression of differentiated functions in hepatoma cell hybrids. VI. Extinction and re-expression of liver alcohol dehydrogenase. *Biochimie* *54*:195–201.
3. Bertolotti, R., and Weiss, M. C. *(in preparation)*: Expression of differentiated functions in hepatoma cell hybrids. V. Re-expression of aldolase B *in vitro* and *in vivo*.
4. Sparkes, R. S., and Weiss, M. C. (1973): Expression of differentiated functions in hepatoma cell hybrids. Alanine aminotransferase. *Proc. Nat. Acad. Sci. USA* *70*:377–381.
5. Weiss, M. C., and Chaplain, M. (1971): Expression of differentiated functions in hepatoma cell hybrids: Reappearance of tyrosine aminotransferase inducibility, after the loss of chromosomes. *Proc. Nat. Acad. Sci. USA* *68*:3026–3030.

DISCUSSION

Siniscalco: Would you say something about the situation with G6PD? I have two of your rat hepatoma lines, the one (FU-5) that was known to be deficient in G6PD and the other (FU-5AH) that has high G6PD activity. I wonder how this type of situation can be explained. Here is an essential function that, to begin with, was absent in FU-5. It then reappeared, presumably as the result of the selection of a subline.

Weiss: This turns out to be a fairly complicated story and was worked on by Jean Deschatrette. We wanted to look at G6PD with the idea of using it as a marker in intraspecific hybrids, and we began systematically to analyze hepatoma clones to verify the absence of this enzyme. Nobody knows why it is absent. However, in some of the subclones of the original hepatoma, we found G6PD was expressed. Interestingly enough, in almost all the cases in which G6PD is expressed, the liver specific enzymes are not. In other words, all of our dedifferentiated clones have G6PD. There are some of the differentiated ones with a trace of it. This seems to be some kind of regulatory phenomenon. Whether it has anything to do with differentiation, I don't know.

To go back to the question of membranes brought up by Attardi, all of the dedifferentiated hepatoma cells are characterized by a different morphology. They spread much more. One of the groups of dedifferentiated hepatoma cells isolated was picked up quite by accident. We have a subline of the hepatoma cells that is resistant to 8-azaguanine and that apparently did not revert. In one cross with fresh liver suspensions, colonies appeared that didn't look like the hepatoma and that were growing next to colonies of fresh liver cells. When these colonies were subcultured in HAT, they died. In other words, as soon as they were in pure culture in HAT, they died. Moreover, they were all "dedifferentiated," and they all had G6PD. It seemed as if there was a correlation in this case between the ability to carry on metabolic cooperation (as these cells did not live by themselves in HAT but could form nests next to other colonies), membrane changes, G6PD expression and the expression of differentiated functions.

Silagi: I want to ask a technical question. When you crossed the nonexpressing hepatoma with the expressing hepatoma, how could you tell that you had hybrids?

Weiss: We had a selective marker in the "dedifferentiated" cells. In addition, there were karyotypic markers.

Thompson: Would you tell us about the quantitative nature of the reexpression of the four different functions?

Weiss: The activity of aldolase B, when reexpressed, is as high as or higher than that of the hepatoma cells. Alcohol dehydrogenase is very difficult to quantitate because the hepatoma cells exhibit two bands. One of them seems to be stomach alcohol dehydrogenase, the other liver alcohol dehydrogenase (see ref. 2). In general the reexpression of alcohol dehydrogenase is intermediate. Alanine aminotransferase is as high as in the hepatoma parental cells, or it can be intermediate. In the first series of reexpressing hybrids (hepatoma-BRL), there was no detectable TAT baseline, and the inducibility was very low compared to the hepatoma cells. The factor of induction was very similar, but the actual induced specific activity was only about one-fiftieth of the hepatoma cells. In the Chinese hamster hybrids, the level of reexpression of TAT is not as high as in the hepatoma cells, but is between 50 and 70%.

Thompson: How about the coordinate reexpression in the hybrids where several things come back together? Do all functions return proportionately?

Weiss: There are only two clones, and the results are too preliminary to allow any firm conclusions. In both cases there is about 30 to 40% of the levels in the hepatoma cells.

Thompson: For all of the functions?

Weiss: We have not yet properly quantitated aldolase B and alcohol dehydrogenase.

Ruddle: Arthur McMorris in our laboratory made hybrids between neuroblastoma cells and mouse fibroblasts, and several differentiated functions were analyzed in the hybrids. One interesting result was that there was a general correlation between the morphology and the number of neuroblastoma traits that were expressed in the hybrids. The results are very much in parallel with the data just presented by Weiss.

We also have some interesting results with regard to what we call "constitutive" functions. In the case of MPI and GOT, we have considerable data that show the assignment of these loci to particular chromosomes. In a few cases, we have observed the presence of the chromosome that is responsible for the production of the phenotype, whereas the phenotype was not expressed. In one case, GOT activity reappeared upon subcloning, showing that the gene actually was there, although not being expressed. Thus, we may find the distinction between the so-called facultative functions and constitutive functions to be a very hazy one, and we may also see regulation of so-called constitutive functions in the hybrids. There may be more to the regulation of these functions than we know about, because, in cases where they are extinguished, the cells do not survive.

Minna: I think one question that has to be raised in all cases of reexpression is whether the line that is exhibiting reexpression or the parental line from which it was derived is truly clonal in nature, i.e., whether they have truly been derived from a single cell. Obviously, if there is a mixed population of cells, 99.99% of which don't express a function and 0.01% of which do, under selective pressure that rare cell that never in its hybrid lifespan had really lost its differentiated state can appear. I think that very strict criteria for cloning should be applied to hybrids to isolate lines that first are extinguished and then to look for reexpression in their progeny.

Ephrussi: Weiss said, in reference to the dedifferentiated hepatoma cells, that these rare variants simultaneously lose all functions. It is not certain that the functions were lost simultaneously, since the cells could have lost the functions independently one after another.

Weiss: I should have said that the cells are characterized by the simultaneous absence of all of the functions, not that they lost them simultaneously.

Somatic Cell Hybridization, edited by
Richard L. Davidson and Felix de la Cruz.
Raven Press, New York © 1974.

Production of Human Albumin in Mouse Hepatoma–Human Leukocyte Hybrids

Gretchen Darlington

Department of Biology, Kline Biology Tower, Yale University, New Haven, Connecticut 06520

We were interested in examining the expression of differentiated functions in hybrids between a mouse hepatoma line and cells of human origin. Our anticipation was that some of the specialized functions of the hepatoma would be extinguished by the human genome. As the human chromosomes were lost from the hybrid, it should then be possible to associate the extinction of a particular hepatic function with a specific human chromosome. Instead, we observed the activation of a human gene, opening up new possibilities for mapping.

We used an HGPRT-deficient line of mouse hepatoma cells called Hepa-1A. The modal chromosome number of the cells is 58 and includes four metacentrics that are distinguishable from the human chromosomes, as well as one long acrocentric chromosome with several heterochromatic regions. The cells express many differentiated functions (see Table 1). These include ES 2 esterase, an enzyme found in large amounts in liver and kidney. (Caution should be exercised in considering ES 2, since it is also present in some mouse fibroblast lines.) Hepa-1A cells do not have more glycogen than fibroblasts (tested with Schiff's reagent) and do not express liver alcohol dehydrogenase, aldolase B or xanthine oxidase (the latter is not liver specific but occurs with high activity in the liver). The cells have several times more tyrosine transaminase (TAT) activity than human fibroblasts and are inducible for TAT by 10^{-5} M dexamethasone. The kinetics of induction of TAT in Hepa-1A cells differ from those of HTC cells (3) or the Reuber hepatoma (2). For full induction, a 48-hr exposure to dexamethasone is necessary. Hepa-1A cells also have high aryl-hydrocarbon hydroxylase activity and this activity is elevated by phenobarbital (a liver specific trait). This enzyme was assayed for us by D. Nebert. The cells also synthesize and secrete mouse serum albumin. An immunoelectrophoretic technique was used for the detection of serum albumin, and both mouse and human albumin can be distinguished and the amount of albumin quantitated. Hepa cells secrete albumin at a rate of 30 n/mg protein/hr. Some 15 subclones of Hepa-1A have been examined, and, whereas the rate of secretion of mouse serum albumin can show as much as a 10-fold varia-

TABLE 1. *Hepatic phenotypes examined in Hepa-1A cells*

Stored glycogen	−
Es-2 esterase	+
Alcohol dehydrogenase	−
Xanthine oxidase	−
Aldolase B	−
Tyrosine aminotransferase	+
TAT inducibility	+
Aryl hydrocarbon hydroxylase	+
AHH inducibility	+
Albumin synthesis	+

tion, no subclone has been found that failed to produce mouse albumin. In fact, the level of albumin secretion of the least active subclones was not lower than that of the uncloned population.

The Hepa-1A cells were fused with human leukocytes from a normal individual and five hybrid clones (designated Hal) were isolated. With the exception of clones 7a and 7b, which were isolated from the same dish, the clones were of independent origin. These hybrids were examined for about 30 enzymes. All the hybrids had at least two human enzymes as demonstrated by the presence of the human band or a mouse-human heteropolymer on starch electrophoretograms. These enzymes are listed in Table 2. The three X-linked markers were present in all the clones except 7a, which did, however, have the human HGPRT. The small number of human enzymes in the hybrids indicated marked segregation of human chromosomes.

The hybrids have been examined for the specialized liver functions. All hybrids expressed the mouse ES 2 activity. None had liver alcohol dehydrogenase or xanthine oxidase, which are lacking in Hepa-1A. Only one clone, 7a, has been tested for TAT activity. The baseline activity was similar to that in human fibroblasts, and the cells were not inducible for TAT with dexamethasone.

All the hybrids produced mouse serum albumin at about the same level as

TABLE 2. *Enzymes examined in hybrid cells*

Cell line	G6PD	HGPRT	PGK	IPO-B	NP	MOD-S	GPI
Hepa-1a	M	−	M	M	M	M	M
Hal 5	MH	H	MH	M	M	M	M
Hal 3	MH	H	MH	M	M	M	MH
Hal 6	MH	H	MH	MH	M	MH	M
Hal 7a	M	H	M	M	MH	MH	M
Hal 7b	MH	H	MH	M	MH	M	M

H—human homopolymer and/or hybrid heteropolymer.
M—mouse homopolymer only.

the mouse hepatoma parent cell. Two of the clones, 7a and 7b, also produced human serum albumin. These hybrids produced about 100 times more mouse serum albumin than human albumin. Thus in these clones, not only was the differentiated mouse function expressed but the corresponding human activity was induced.

The chromosome complements of clones 3, 6, and 7a have been examined. The modal chromosome numbers of clones 3 and 6 are similar to that of Hepa-1A. Hal 3 and 6 have modal numbers of 59 and 60, respectively. We were most interested in the hybrids which produced human albumin. Hal 7a appears to be a 2S hybrid with a modal chromosome number of 106. On the basis of chromosome-banding techniques, Hal 3 has retained the human X chromosome in at least 36% of the cells. Hal 6 has the human X and chromosomes 6 and 8. The retention of chromosome 6 is confirmed by the isozyme data. In Hal 7a, no intact human chromosome has been detected, nor can any human chromosome fragment translocated onto a mouse chromosome be recognized. The chromosome analysis of Hal 7b is in progress.

Our results indicate that the genome of a cell type that does not ordinarily produce serum albumin can be induced to do so under certain conditions (*cf.* ref. 1). At least for serum-albumin synthesis, the human parent cell need not be a hepatoma cell in order to have albumin synthesis in the hybrids. We anticipate that the human chromosome carrying the gene for human serum albumin can soon be identified in the hybrid cells.

REFERENCES

1. Peterson, J., and Weiss, M. (1972): Expression of differentiated functions in hepatoma cell hybrids: Induction of mouse albumin production in rat hepatoma–mouse fibroblast hybrids. *Proc. Nat. Acad. Sci. 69*:571.
2. Schneider, J., and Weiss, M. (1971): Expression of differentiated functions in hepatoma cell hybrids, I. Tyrosine aminotransferase in hepatoma-fibroblast hybrids. *Proc. Nat. Acad. Sci. 68*:127.
3. Tomkins, G., Thompson, E., Hayashi, S., Gelehrter, T., Granner, D., and Peterkofsky, B. (1966): Tyrosine transaminase induction in mammalian cells in tissue culture. *Cold Spr. Harb. Symp. Quant. Biol. 31*:349.

DISCUSSION

Gerald: Could you give a summary of your selection for these clones?

Darlington: The Hepa-1A parent is HGPRT deficient. The leukocytes do not attach to the substrate. Thus, we could use the half selective system and simply wash the leukocytes away.

Gerald: Did I understand that clone 7a has human HGPRT?

Darlington: Yes, although it does not have human G6PD or human PGK.

Davidson: Have you isolated any subclones from clone 7a which have lost the human albumin so that you could try to make a correlation with the loss of some other human enzymes?

Darlington: Preliminary results suggest a possible association between the loss of human albumin and the loss of nucleoside phosphorylase. This is extremely tentative, however. These subclones seem to be reduced for everything, and it is therefore difficult to say whether the loss of these two functions is related or not.

Davidson: Is the nucleoside phosphorylase activity present in any of the hybrids which do not make albumin?

Darlington: No.

Weiss: When you say that some of the subclones of this hybrid 7a do not produce albumin, do you mean they don't produce human albumin or they don't produce any albumin at all.

Darlington: They do not produce human albumin. They do produce mouse serum albumin. We have never observed in any of the subclones of 7a the loss of mouse albumin.

Pontecorvo: Just a matter of biological safety: it would be advisable not to use lymphocytes from anyone who comes near the lab at all for this sort of work.

Darlington: I agree with the warning.

Somatic Cell Hybridization, edited by
Richard L. Davidson and Felix de la Cruz.

Unexpected Production of a Mouse Serum Protein by Rat-Mouse Somatic Cell Hybrids

Priscilla S. Dannies* and Armen H. Tashjian, Jr.

Laboratory of Pharmacology, Harvard School of Dental Medicine and Department of Pharmacology, Harvard Medical School, Boston, Massachusetts 02115

INTRODUCTION

We have hybridized the mouse fibroblast line C1-1D with a serum albumin-producing rat hepatoma cell strain and have observed the production by some of the hybrid clones of a mouse serum protein that is not albumin. This brief report describes the results of those experiments.

MATERIALS AND METHODS

Cell Lines and Their Characteristics

The mouse cells were clone 1D (C1-1D), a fibroblast line derived from L cells that is deficient in thymidine kinase (3). C1-1D does not produce mouse serum albumin or other mouse serum proteins as measured by our complement fixation assay method (see below). The rat cells were a strain developed in our laboratory from Reuber H-35 hepatoma cells (H_4 II EC_3, see ref. 7) by selection for their ability to grow in 8-azaguanine (25 μg/ml). These hepatoma cells were designated H_4AzC_2. They produce rat serum albumin and contain a hydrocortisone-stimulated tyrosine aminotransferase.

Cells were grown in Ham's F 10 medium supplemented with 15% horse serum and 2.5% fetal calf serum. They were cultured at 37°C in plastic Petri dishes (Falcon) in a humidified atmosphere of 5% CO_2 and 95% air.

Hybridizations of H_4AzC_2 and clone 1D were done using lysolecithin (1). Lysolecithin from Sigma was prepared as described (1) and was used at a final concentration of 125 μg lysolecithin/ml in a solution of 5 mg/ml bovine serum albumin in phosphate buffered saline. Hybrid cells were selected by plating the fused cells in HAT medium. After 2 weeks, four colonies were selected, HD_1, HD_2, HD_3, and HD_4. These were cloned 2 weeks later, giving the subclones described. Karyotypes of cells were prepared according to the method of Tijo and Wang (9).

* Research fellow of The Arthritis Foundation.

Immunoassay Methods for Rat Serum Albumin and Mouse Serum Albumin

Rat serum albumin (RSA) was measured by microcomplement (C′) fixation using rabbit antiserum (8). Under the conditions used, mouse albumin did not cross-react significantly in the assay for rat albumin.

Antiserum against mouse serum albumin (MSA) (Pentex Fraction V Powder Lot 10) was prepared in a rabbit. Quantitative C′ fixation (4) was used to measure the mouse albumin–antimouse albumin reaction. The dilution of rabbit anti-MSA used to measure the homologous antigen was 1:10,000. At this dilution, RSA did not cross-react in the mouse albumin immune system.

Results are expressed as micrograms albumin produced per milligram cell protein per 24 hr or micrograms albumin per milliliter medium. Cell protein was determined by the method of Lowry et al. (5).

Immunoelectrophoresis

Immunoelectrophoresis buffer was 0.05 M Tris-barbital buffer, pH 8.8. The samples were subjected to electrophoresis in 1% Difco Noble Agar on slides at 2 mA/cm for about 2 hr.

Measurement of Tyrosine Aminotransferase Activity (TAT)

TAT activity was determined by the method of Diamondstone (2). The results are given as micromoles of product (*p*-hydroxybenzaldehyde) formed per milligram cell protein per minute.

RESULTS AND DISCUSSION

TAT Activity

Basal TAT activity was 4.5 μmoles/mg protein/min in the rat cell strain, and the activity was increased six- to 10-fold by incubation of the cells with hydrocortisone at 10^{-6} M for 24 hr. The mouse cells had low or unmeasurable TAT activity (<0.3 to 1.0 μmoles/mg protein/min), and the activity was not enhanced by incubation with hydrocortisone. The hybrid cell strains behaved like the mouse cell parent.

Serum Albumin Production

Production, in this context, is defined as the accumulation of albumin in the culture medium where it is not degraded (8). The rat parental strain produced substantial quantities of RSA while C1-1D produced no detectable MSA or RSA (Table 1). Two of the primary hybrid clones (HD_1 and

TABLE 1. *Production of rat and mouse serum albumin by parental and hybrid clones*

Cell type	Albumin production[a] RSA (μg/mg protein/24 hr)	MSA (μg/mg protein/24 hr)
H_4AzC_2	12–18	nd[b]
C1-1D	nd	nd
HD_1	nd	—
HD_2	nd	nd
HD_3	0.02	nd
HD_4	0.06	nd
HD_4C_1	1.0	nd
HD_4C_2	1.0	nd

[a] Each value is the mean of at least two determinations from two separate medium collections.
[b] None detected, <0.005 μg/mg protein/24 hr.

HD_2) produced no detectable albumin (Table 1); however, the other two clones (HD_3 and HD_4) did produce small amounts of RSA, and two subclones of HD_4 (HD_4C_1 and HD_4C_2) produced more RSA than the parental hybrid clone (Table 1). None of the clones produced measurable quantities of MSA.

In order to demonstrate whether the albumin produced by the hybrid clones was of rat or mouse origin, the medium was tested by micro-C′ fixation against anti-RSA and anti-MSA at the dilutions of each antiserum used to measure the homologous antigen. The results obtained with anti-RSA are presented in Fig. 1. These data show that the antigen being measured in the medium of HD_4C_1 and HD_4C_2 cells gives C′-fixation curves which are essentially the same as that obtained with RSA. In particular, the amount of C′ fixed at antigen-antibody equivalence is the same for all three C′-fixation curves (Fig. 1). This C′-fixation method can detect small differences in antigenic structure, which are reflected in a decrease in the amount of C′ fixed in the zone of equivalence (4). At a dilution of anti-RSA of 1:1800, MSA did not fix C′ at all with this antiserum.

The results obtained with anti-MSA are shown in Fig. 2. The optimum dilution of anti-MSA used to measure the homologous antigen was 1:10,000. At this dilution of anti-MSA, a small amount (10%) of C′ was fixed with medium from HD_4C_1 and HD_3 cell lines. In order to accentuate the cross-reactions obtained with media from these cells, we carried out the C′-fixation reaction with anti-MSA diluted 1:6000. As expected, MSA gives a flat extended zone of antigen-antibody equivalence (Fig. 2). Medium from HD_3 and HD_4C_1 cells gives complete C′-fixation curves, but the amount of C′ fixed at equivalence was less than that fixed with MSA as the antigen, which

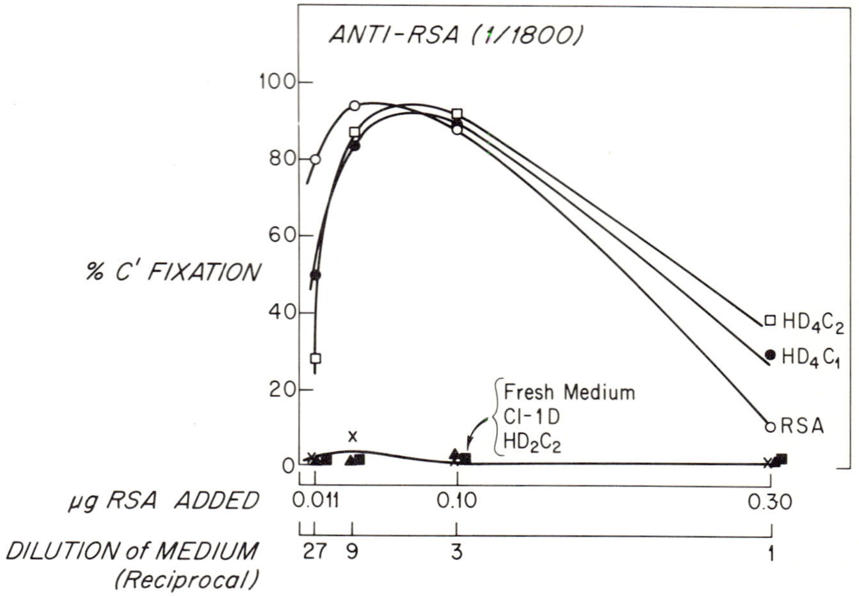

FIG. 1. C′-fixation curves obtained with anti-RSA diluted 1/1800 and RSA, media from several cell lines, or fresh control medium. The initial dilutions of fresh medium, C1-1D and HD_2C_2 media were 1:3, whereas the initial dilutions of HD_4C_1 and HD_4C_2 were 1:5.

demonstrates that the antigen present in the medium from hybrid cells is not MSA. The presence of the cross-reacting antigen in hybrid medium was illustrated more convincingly if the anti-MSA was diluted only 1:2000 (Fig. 3). At dilutions of anti-MSA of 1:2000 and 1:6000, small cross-reactions were seen with RSA itself and RSA produced by H_4AzC_2 cells, whereas no C′ was fixed with fresh medium or medium from C1-1D and HD_4C_2 cells (Figs. 2 and 3).

We have called the antigen measured in medium of HD_3 and HD_4 (data not shown) and HD_4C_1 cells mouse serum protein (MSP) to differentiate it from MSA. It is a mouse serum protein (detected in mouse serum by C′ fixation, data not shown) which must have been a contaminant in the MSA preparation used for immunization. This contaminant, which we estimate by C′ fixation was present to an extent of about 1% in the MSA preparation, elicited specific antibody formation in the rabbit immunized with MSA. This second immune system (MSP–anti-MSP) was detected at a dilution of anti-MSA of 1:6000 or less, but not significantly at the dilution (1:10,000) used to measure MSA.

Further evidence that the MSP measured with anti-MSA is not MSA was obtained by immunoelectrophoresis (Fig. 4). Whole mouse serum was subjected to electrophoresis, and the separated proteins were allowed to react

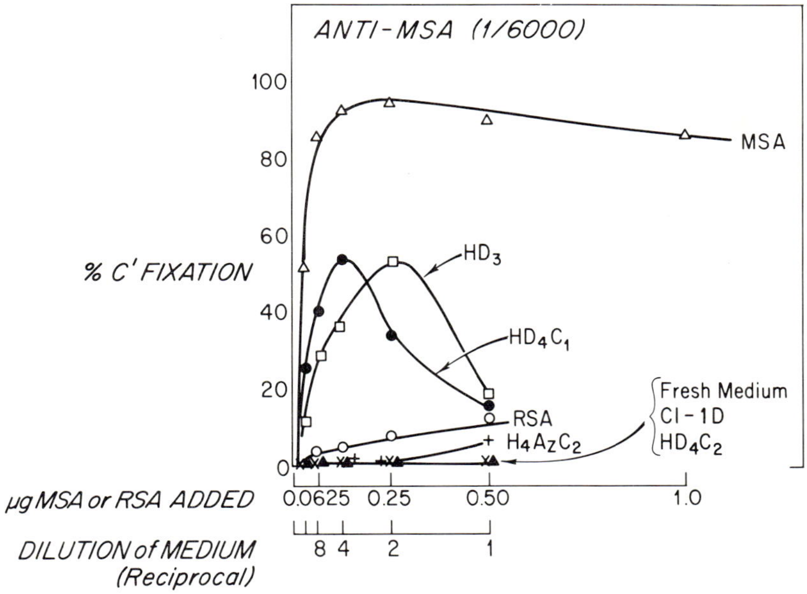

FIG. 2. C′-fixation curves obtained with anti-MSA diluted 1/6000 and MSA, RSA, and the various indicated cell culture media. The initial dilutions of fresh medium and media from C1-1D, HD_4C_2, and H_4AzC_2 were 1:3, whereas the initial dilutions of HD_3 and HD_4C_1 were 1:5.

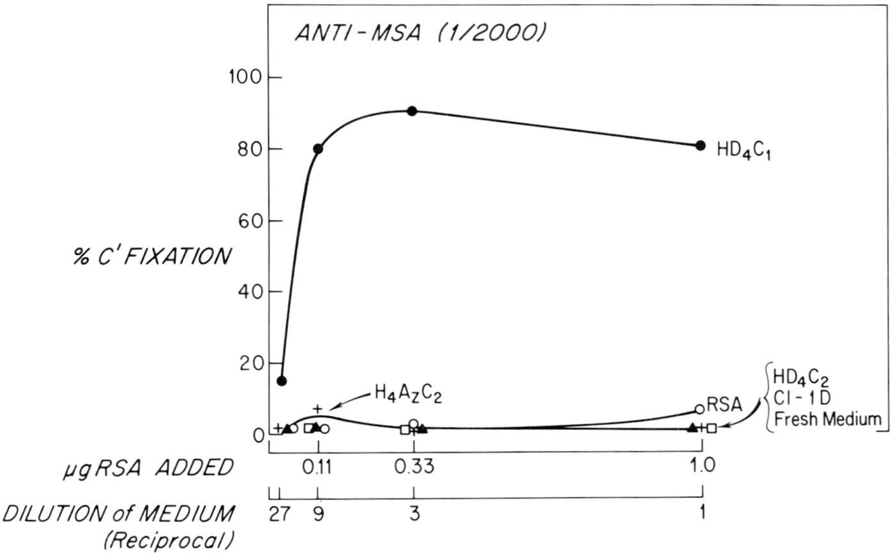

FIG. 3. C′-fixation curves obtained with anti-MSA diluted 1/2000 and RSA and various cell culture media diluted as described in the legend for Fig. 2.

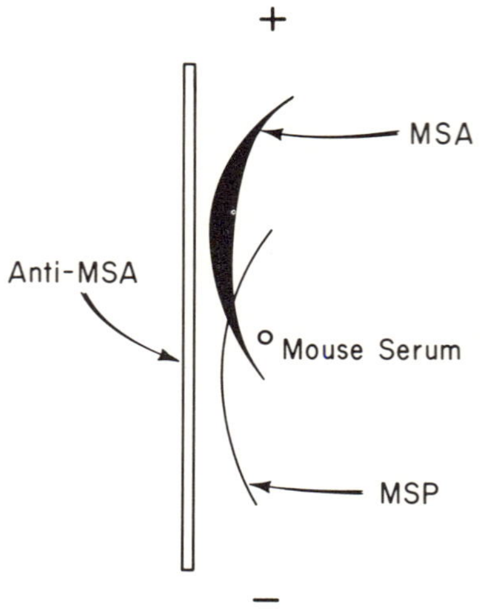

FIG. 4. Immunoelectrophoresis of whole mouse serum and anti-MSA.

with anti-MSA. Two bands of precipitation were observed: one dense band moved toward the anode and was MSA; the second band moved toward the cathode and is designated MSP (Fig. 4). We do not know which protein of mouse serum is represented by MSP. If concentrated medium from HD_4 or HD_4C_1 cells was examined by immunoelectrophoresis, a weakly staining, cathodally migrating band was seen with a mobility essentially the same as that observed in whole mouse serum.

Taken together, these data lead us to the following conclusions. (a) Two primary (HD_3 and HD_4) and two secondary (HD_4C_1 and HD_4C_2) hybrid clones produce RSA, but the production of RSA by these hybrids is less than that of the H_4AzC_2 parent strain. (b) Three primary (HD_2, HD_3, and HD_4) and one secondary hybrid clones produce a protein found in mouse serum that is not produced in detectable amounts by either parental cell line. It is possible that the H_4AzC_2 rat cell strain produces an analogous rat serum protein.

A summary of the characteristics of these hybrid cell strains is shown in Fig. 5. The modal chromosome numbers determined by us in H_4AzC_2 and C1-1D were 46 and 50, respectively. The primary hybrid clones (HD_1, HD_2, HD_3, and HD_4) contained 89 to 96 chromosomes. The production or lack of production of RSA, MSA, and MSP in these primary hybrids is shown in Fig. 5. The secondary clones showed chromosomal loss with numbers of chromosomes varying from 60 to 89 per cell. Two of these subclones (HD_4C_1 and HD_4C_2) produced more RSA than the primary line from which they were derived. One of the secondary clones (HD_4C_1) also produced

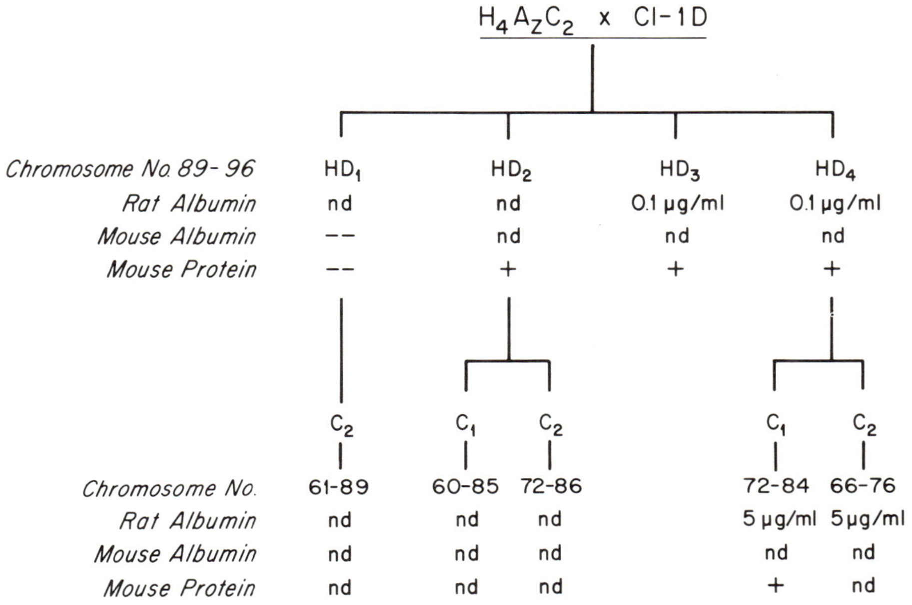

FIG. 5. Summary of karyotype data and phenotypic expression of various primary and secondary hybrid clones produced by fusion of H_4AzC_2 and C1-1D cells. The amounts of MSP (mouse protein) produced cannot be assigned a numerical value because we do not have a primary standard for this material to use in the C′-fixation assay. We have, therefore, designated its presence by +, and its absence by nd (none detected).

MSP (Fig. 5). It may be significant that HD_4C_1, which produced MSP, had on the average more chromosomes than HD_4C_2, which did not produce MSP. We did not, however, determine whether the extra chromosomes were of rat or mouse origin.

The results of these experiments reveal that the synthesis and secretion of RSA is not uniformly extinguished in hybrid cells formed between albumin-producing rat hepatoma cells and mouse fibroblasts. We obtained no evidence for the "induction" of MSA production in the hybrid cells (6). A more novel finding, however, was the synthesis and secretion of a mouse serum protein, which was not albumin, by some of the hybrid clones. This protein was not synthesized by the mouse parental line. If the analogous protein was also not synthesized by the rat parent, these results would then be an example of the expression of a phenotype in hybrid cells that was not characteristic of either parent line. If, on the other hand, H_4AzC_2 cells did produce a rat protein analogous to MSP, our results show that a less than tetraploid dose of rat genetic material was able to induce the production of the analogous mouse protein in the hybrid cells.

ACKNOWLEDGMENTS

The authors wish to thank Misses Yolanda Santo and Diane Jensen for expert assistance and Dr. U. I. Richardson for help and critical advice.

This investigation was supported in part by a research grant from the National Institute of Arthritis, Metabolism and Digestive Diseases (AM 11011).

REFERENCES

1. Croce, C. M., Sawicki, W., Kritchevsky, D., and Koprowski, H. (1971): Induction of homokaryocyte, heterokaryocyte and hybrid formation by lysolecithin. *Exp. Cell. Res. 67*:427–433.
2. Diamondstone, T. I. (1966): Assay of tyrosine transaminase activity by conversion of *p*-hydroxyphenylpyruvate to *p*-hydroxybenzaldehyde. *Analyt. Biochem. 16*:395–401.
3. Hsu, T. C., and Merchant, D. J. (1961): Mammalian chromosomes *in vitro* XIV. Genotypic replacement in cell populations. *J. Nat. Cancer Inst. 26*:1075–1081.
4. Levine, L. (1968): Micro-complement fixation. In: *Handbook of Experimental Immunology* (ed. D. M. Weir), pp. 707–716, Blackwell Sci. Publ., Oxford.
5. Lowry, O. H., Rosebrough, N. J., Farr, A. L., and Randall, R. J. (1951): Protein measurement with the folin phenol reagent. *J. Biol. Chem. 193*:265–275.
6. Peterson, J. A., and Weiss, M. (1972): Expression of differentiated functions in hepatoma cell hybrids: Induction of mouse albumin production in rat hepatoma–mouse fibroblast hybrids. *Proc. Nat. Acad. Sci. 69*:571–575.
7. Pitot, H. C., Peraine, C., Morse, P. A., and Potter, V. R. (1964): Hepatomas in tissue culture compared with adapting liver *in vivo*. *Nat. Cancer Inst. Monograph 13*:229–241.
8. Richardson, U. I., Tashjian, A. H., Jr., and Levine, L. (1969): Establishment of a clonal strain of hepatoma cells which secrete albumin. *J. Cell Biol. 40*:236–247.
9. Tijo, J. H., and Wang, J. (1965): Direct chromosome preparations of bone marrow cells. In: *Human Chromosome Methodology* (ed. J. Y. Yvuis), pp. 51–56. Academic Press, New York.

DISCUSSION

Weiss: Would you summarize the evidence for saying that the new mouse serum protein does not have an equivalent protein that is produced by the rat hepatoma cells? The Reuber cells produce a wide variety of serum proteins.

Tashjian: We do not find any measurable MSP in the medium of H_4AzC_2 cells or in the cells even using dilutions of anti-MSA of 1:2000. However, the small degree of cross-reaction between anti-MSA and RSA makes it impossible to rule out the presence of a small amount of MSP-like material in the RSA immunogen or in H_4AzC_2 cells or medium.

Weiss: You would only pick up a cross-reacting protein, since this is a contaminant that was in the mouse serum albumin preparation and perhaps was not in the rat serum albumin preparation.

Tashjian: That is possible.

Gerald: Did you test the antibody against whole rat serum?

Tashjian: Yes. A MSP-like material cannot be detected in whole rat serum.

Gerald: I am raising the same objection Weiss did. When you say neither parent produces this protein, a homologous protein produced by the rat hepatoma but not cross-reacting with the antirat albumin antibody would not be detected. Weiss is perhaps thinking of the situation of the rat albumin-producing hepatoma stimulating production of mouse albumin. Possibly your hepatoma line is producing something that you have as yet not measured or detected, but the homologous mouse protein is stimulated in the hybrids and is detected because of the particular antibody used.

Tashjian: That is possible.

Siniscalco: What about the possibility of a hybrid molecule of some sort, part of which already existed in the hepatoma cells?

Tashjian: This is certainly possible. However, if an analogous molecule existed in the rat hepatoma cells, there is a reasonable likelihood that it would have been detected using a high concentration of anti-MSA and the culture medium or extracts of H_4AzC_2 cells. In other words, if part of the antigenic determinant were present in the rat parental cells, one might expect to detect it. This, however, is not 100% certain. We have looked for the ability of rat hepatoma cell media and cell extracts to inhibit the antigen-antibody reaction obtained between the MSP and anti-MSA and have seen no inhibition. These data also help to rule out the possibility that there was an analogous protein of rat origin. However, a negative result is not firm evidence.

Minna: I wonder if extracts of the cells incubated with either rat albumin or rat serum could yield some degradative product that could give this reaction. Have you done co-cultivation experiments or mixing of extracts?

Tashjian: We have not made mixtures of the extracts or performed co-cultivation experiments.

Gerald: On the same point, is the electrophoretic mobility of the protein present in the medium from the hybrid cells equivalent to that seen in the whole mouse serum?

Tashjian: That is difficult to say, because one has to concentrate the medium in order to see MSP by electrophoresis. The complement-fixation procedure is much more sensitive than the electrophoretic procedure. We have not yet run the hybrid cell medium and whole mouse serum on the same gel to determine whether or not there is identity between the protein in the two samples.

Gordon: Do you know if different batches of albumin have the same contaminant? There are many enzymes, for example, like β-galactosidase or phospholipase, in highly purified crystallized preparations from most commercial sources.

Tashjian: We have not looked at multiple commercial sources. Certainly, the original material is significantly heterogeneous.

Somatic Cell Hybridization, edited by
Richard L. Davidson and Felix de la Cruz.
Raven Press, New York © 1974.

Regulation of the Inducibility of Tyrosine Aminotransferase in Rat-Human Hybrids

Carlo M. Croce, Gerald Litwack, and Hilary Koprowski

Wistar Institute of Anatomy and Biology, 36th Street at Spruce, Philadelphia, Pennsylvania 19104 and Fels Research Institute, Department of Biochemistry, Temple University School of Medicine, Philadelphia, Pennsylvania

The inducibility of the enzyme tyrosine aminotransferase (TAT) by dexamethasone was studied in hybrids between inducible rat hepatoma cells (Fu5AH) and noninducible human fibroblasts (WI-38) (3).

The rat hepatoma parental cells, deficient in hypoxanthine guanine phosphoribosyltransferase (HGPRT), were fused with the HGPRT-positive WI-38 fibroblasts (5) in the presence of β-propiolactone-inactivated Sendai virus (2). The hybrid cell colonies selected in HAT medium (6) were picked, and then the hybrid cells were cloned.

The rat-human hybrid cells, similar to other rodent-human hybrids (1, 7, 8), rapidly lost the majority of human chromosomes (Fig. 1). The human chromosome, which carries the human HGPRT gene (4), was maintained in the hybrid clones selected in HAT medium. When the clones were back-selected in medium containing 8-azaguanine, it was possible to obtain clones that had lost the human X-chromosome. The rat hepatoma cells contained adenine phosphoribosyltransferase (APRT) and glucose-6-phosphate dehydrogenase (G6PD). The hybrids selected in HAT medium displayed rat or rat and human APRT, human HGPRT, and rat and human G6PD (Table 1). The hybrid clones back-selected in a medium containing 8-azaguanine were found to have APRT and rat G6PD activities but not HGPRT (Table 1).

We tested the rat and human parental cells and the hybrid clones for inducibility of TAT by dexamethasone and, in some hybrid clones, we also studied the effect of actinomycin D on the inducibility of TAT in the parental and hybrid cells.

In the hybrid clones selected in HAT medium that contained the human X-chromosome and displayed human HGPRT and G6PD activities, TAT was not induced by 2×10^{-5} M dexamethasone (Table 1). In all the clones back-selected in 8-azaguanine that did not contain the human X-chromosomes and human HGPRT and G6PD activities, and, in the parental rat hepatoma cells, TAT was inducible by dexamethasone (Table 1). For example, in clone 1 (HAT) TAT was not inducible, whereas in the same clone

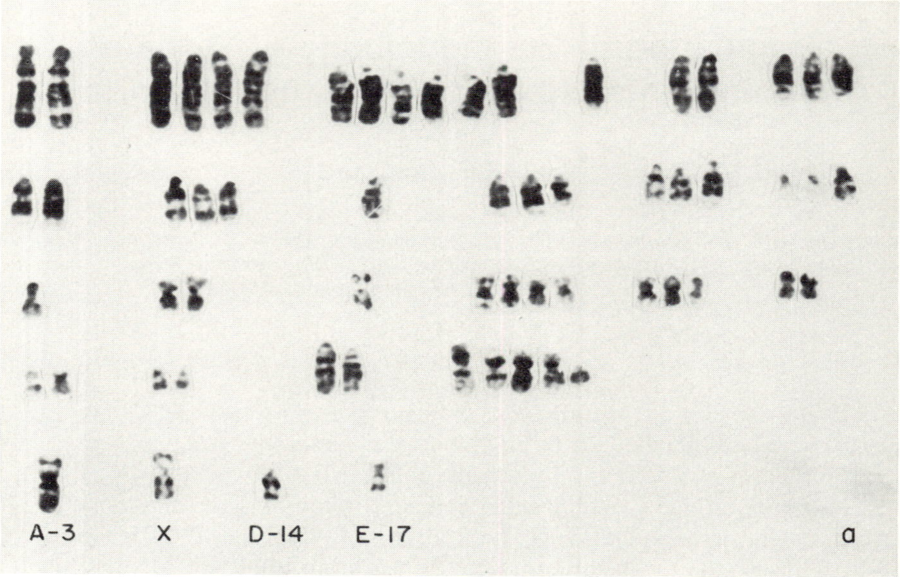

FIG. 1. Karyotype of a rat-human hybrid clone. The human chromosomes A-3, X, D-14, and E-17 are present in this clone. All the other chromosomes are of rat origin.

back-selected in 8-azaguanine TAT was inducible. To determine if this noninducibility of TAT was due to an absence, or a decrease, of the receptor activity for dexamethasone in the hybrid cells containing the human X-chromosome, we measured the corticosteroid receptor activity in the parental and in the hybrid cells.

The dexamethasone receptor(s) were expressed to the same extent in the suppressed as in the inducible hybrid clones (Table 2). In additional experiments, we exposed the parental cells and the suppressed and inducible hybrid clones to concentrations of dexamethasone varying from 10^{-4} to 10^{-8} molar for 18 hr. After that, we replaced the medium of the flasks with medium containing 5 μg/ml of actinomycin D with and without dexamethasone for 6 hr. To our surprise we found that the superinducibility of TAT by actinomycin D in the rat parental cells and in the 8-azaguanine clones did not exceed 20% of the inducible level. In three suppressed clones, actinomycin D was able to induce TAT, but one suppressed clone failed to show inducibility of TAT by actinomycin D.

These results lend support to the hypothesis of a regulatory function of the human X-chromosome on the inducibility of tyrosine aminotransferase by corticosteroid hormones in rat-human hybrids. Whether this suppression involves a human regulatory protein that is able to shut off the inducibility of TAT at either transcriptional or post-transcriptional level is still unknown.

TABLE 1. *Induction of TAT and presence of X-linked human isozymes in human-rat hybrid cells*

Cell line		Selective medium	Addition to culture medium	TAT specific activity[a]	X-Linked human isozymes G6PD	X-Linked human isozymes HGPRT
Parental	WI-38	None	None	0	+	+
			Dex[b]	0		
	Fu5AH	8-Aza[c]	None	0.5	−	−
			Dex	1.9		
Hybrid[d]	Cl.1	HAT	None	0	+	+
			Dex	0		
		8-Aza	None	0.3	−	−
			Dex	5.3		
	Cl.2	HAT	None	0	+	+
			Dex	0		
		8-Aza	None	0.7	−	−
			Dex	3.5		
	Cl.3	HAT	None	0	+	+
			Dex	0		
	Cl.4	HAT	None	0	+	+
			Dex	0		
	Cl.21	HAT	None	0.1	+	+
			Dex	0.1		
		8-Aza	None	1.5	−	−
			Dex	6.5		
	Cl.30	HAT	None	0	+	+
			Dex	0		
	Cl.33	HAT	None	0.4	+	+
			Dex	0.5		
	Cl.35	HAT	None	0.6	+	+
			Dex	0.5		
	Cl.8	8-Aza	None	0.7	−	−
			Dex	2.5		
	Cl.11	8-Aza	None	0.4	−	−
			Dex	2.5		
	Cl.19	8-Aza	None	1.6	−	−
			Dex	2.3		
	Cl.24	8-Aza	None	0.6	−	−
			Dex	1.7		
	Cl.25	8-Aza	None	0.6	−	−
			Dex	2.5		
	Cl.28	8-Aza	None	1.0	−	−
			Dex	3.1		
	Cl.29[e]	8-Aza	None	0.2	+	−
			Dex	0.8		

[a] TAT specific activity = μmoles *p*-hydroxy-phenyl pyruvate formed per 10 min (37°C) per mg protein.
[b] Dex = dexamethasone.
[c] 8-Aza = 8-azaguanine.
[d] C1. refers to the clones of WI-38 × Fu5AH hybrid cells.
[e] Human X-chromosome absent; for human karyotype of other hybrids, see Fig. 1 and Table 2.

TABLE 2. *Corticosteroid receptor activity of parental and hybrid cell lines*

Cell line	Presence of human X-chromosome	Receptor activity (% hormone binding)	Ave[a]	TAT inducibility of cell line
WI-38	+	2[b]		−
Fu5AH	−	13		++
Cl.1 HAT[c]	+	10		
Cl.30 HAT[c]	+	18	12	−
Cl.33 HAT[c]	+	9		
Cl.8 8-Aza[d]	−	7		
Cl.19 8-Aza[d]	−	15	12	++++
Cl.25 8-Aza[d]	−	13		

[a] Ave = average values.
[b] Indicates very low or negligible amount of receptor.
[c] HAT = selected in HAT medium.
[d] 8-Aza = selected in 8-azaguanine medium.

ACKNOWLEDGMENTS

This work was supported in part by Public Health Service Research Grants CA 04534 and CA 10815 from the National Cancer Institute, and RR-05540 from the Division of Research Resources.

REFERENCES

1. Croce, C. M., Kieba, I., and Koprowski, H. (1973): Unidirectional loss of human chromosomes in rat-human hybrids. *Exp. Cell Res. 79*:461–463.
2. Croce, C. M., Koprowski, H., and Eagle, H. (1972): Effect of environmental pH on the efficiency of cellular hybridization. *Proc. Nat. Acad. Sci. USA 69*:1953–1956.
3. Croce, C. M., Litwack, G., and Koprowski, H. (1973): Human regulatory gene for inducible tyrosine aminotransferase in rat-human hybrids. *Proc. Nat. Acad. Sci. USA 70*:1268–1272.
4. Grzeschik, K. H., Allderdice, P. W., Grzeschik, A., Opitz, J. M., Miller, O. J., and Siniscalco, M. (1972): Cytological mapping of human X-linked genes by use of somatic cell hybrids involving an X-autosome translocation. *Proc. Nat. Acad. Sci. USA 69*:69–73.
5. Hayflick, L. (1965): The limited *in vitro* lifetime of human diploid cell strains. *Exp. Cell Res. 37*:614–636.
6. Littlefield, J. W. (1966): The use of drug resistant markers to study the hybridization of mouse fibroblasts. *Exp. Cell Res. 41*:190–196.
7. Weiss, M. C., and Green, H. (1967): Human-mouse hybrid cell lines containing partial complements of human chromosomes and functioning human genes. *Proc. Nat. Acad. Sci. USA 58*:1104–1111.
8. Westerveld, A., Meera Khan, P., Visser, R. P. L. S., and Bootsma, D. (1971): Loss of human genetic markers in man–Chinese hamster somatic cell hybrids. *Nature New Biol. 234*:20–24.

DISCUSSION

Davidson: I have two questions. In the hybrids that have an X-chromosome, even though they have a dexamethasone receptor, are you certain that dexamethasone enters the cell? Secondly, in those hybrids that are inducible, is it just the hepatoma enzyme that is induced or is the human enzyme also induced?

Croce: We studied the nuclear transfer of the dexamethasone-receptor complex in at least a few clones. There are no differences between the clones selected in HAT and the clones back-selected in 8-azaguanine. In regard to the second question, we do not yet have the results on the activation of the human form of TAT.

Weiss: I am confused about the superinducibility results. You say that the clones that do not express inducibility with dexamethasone are super-inducible.

Croce: I am talking about superinducibility, when the specific activity of TAT after induction with dexamethasone and then exposure to actinomycin D is higher than after induction with dexamethasone.

Weiss: The initial phenomenon of superinducibility by actinomycin was essentially about a 20% increase over the induced level, is that correct?

Thompson: The degree of superinduction varies greatly.

Weiss: You indicated that the hybrids with an X-chromosome have zero TAT activity.

Croce: In some clones, yes.

Weiss: How much increase in activity do you get with actinomycin?

Croce: We get very low activity after superinduction (about 0.4) in these clones with zero-baseline activity and zero inducibility. In contrast, and I cannot explain why, in those clones with baseline activity of TAT, we obtained higher superinducibility.

Thompson: I will offer an explanation. In general, the absolute level of enzyme that is reached after actinomycin treatment, when the phenomenon occurs, is related to the amount of enzyme activity that one begins with. Why that is so is not in any way clear. The fact that there is less than full return of enzyme activity after actinomycin is thus not surprising.

Weiss: The point that confused me is that, in the work with HTC cells, when there was superinducibility, it was really a very small percent increase over either the baseline or the induced level. Therefore, it looks like this is a different phenomenon.

Croce: If I remember correctly, wasn't there an increase of two- or three fold after actinomycin treatment of HTC cells in your experiments, Dr. Thompson?

Thompson: I don't know whether the percent increase is a valuable way of looking at the data. The percent increase is infinite in hybrids with a baseline of zero. In glutamine synthetase induction, for example, one can induce from no activity or very little enzyme activity to a high activity with actinomycin or other drugs. At the moment, one cannot propose a mechanism which would explain everything with a single model.

Ruddle: Can you detect albumin in the hybrids?

Croce: I am checking that now.

Minna: Do I understand your data correctly; are you assigning a repressor to the X-chromosome?

Croce: Yes.

Somatic Cell Hybridization, edited by
Richard L. Davidson and Felix de la Cruz.
Raven Press, New York © 1974.

Enzyme Expression and Growth Control in HTC Cells and Hybrids of HTC Cells with Other Cell Types

E. Brad Thompson

Laboratory of Biochemistry, National Cancer Institute, National Institutes of Health, Bethesda, Maryland 20014

We have been studying phenotypic variation as it pertains to tyrosine aminotransferase (TAT) inducibility. We began with the HTC line derived from Morris hepatoma 7288C and cloned it under nonselective conditions. When the subclones were analyzed, a wide range of inducibility was found compared to wild type (1). In testing for inducibility, we were very careful to test all the clones at the same time and under the same conditions, because the cells do show day to day variations that are hard to control. Figure 1 shows the results of an experiment in which a set of randomly isolated clones was analyzed for TAT. The induction (basal enzyme activity divided into induced enzyme activity) ranges from very little to 30-fold in these clones. A clone that was highly inducible and one that was only slightly inducible were subcloned, and the secondary clones were analyzed. Again there was a great deal of variation in inducibility among the clones, with considerable overlap between the two sets.

This experiment has been done many times with similar results. Despite the overlapping inducibility of subclones, we find that low-inducing and high-inducing lines generally maintain their widely differing TAT inductions over months of cultivation. We have now looked at some other characteristics of the cells as well as inducibility. The cytosol steroid receptor protein required for TAT induction is present in normal amounts in all the clones tested so far, both inducible and noninducible ones. Thus the latter must be blocked at some subsequent step. There is no correlation between inducibility and basal-TAT level, growth rate, or general protein synthesis. Glutamic-oxaloacetic transaminase, another enzyme that is readily assayed in these cells, appears not to show the same variation in the subclones.

In a less extensive series of experiments, involving about 20 clones, there was a 10-fold variation among clones in the activity of the "detoxifying" enzyme glucuronyl transferase, which exists in the membrane fraction of the cells.

These results suggest that one should be cautious in interpreting experi-

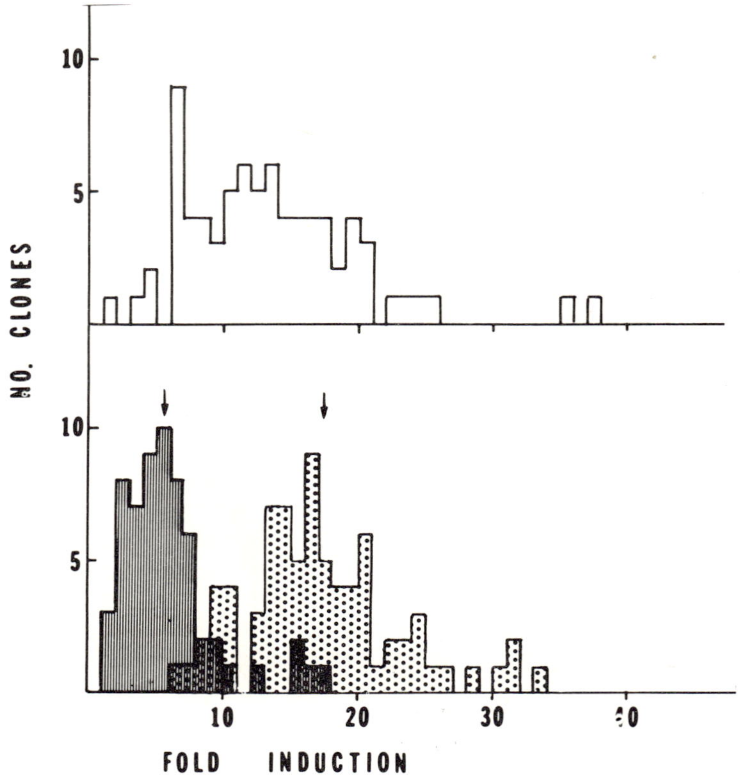

FIG. 1. Distribution of HTC subclones relative to their TAT inducibility by dexamethasone phosphate. Enzyme induction is expressed as the ratio of specific activity in induced cells to that in noninduced cells. Distribution of HTC+ subclones is given at the top, and distributions of C2-10(L) subclones (shaded area) and of C3-30(H) subclones (dotted area) are shown at the bottom. The arrows indicate the degree of induction of C2-10(L) and C3-30(H) clones. (From ref. 1.)

ments in which people look for the return of a function in hybrids in which there has been repression of enzyme inducibility. Weiss and her colleagues have documented the return of TAT inducibility in 3T3-Reuber hepatoma hybrids and have found that inducibility returns at different levels, sometimes far below the original inducible parent (4). One interpretation of those experiments might be that a quantum effect on the inducibility at the regulatory level has occurred. Another possibility is that there is always 100% return of inducibility, but that different hybrids arose from fusions with hepatoma cells with different degrees of inducibility.

In other experiments in our laboratory, 3T3 cells and HTC cells were fused in an attempt to look at the problem mentioned by Davidson, the question of reciprocal control of an inducible function in each parental cell line (2). We collaborated in this work with D. Nebert, who had shown that aryl hydrocarbon hydroxylase is inducible in 3T3 cells by substances such as benzanthracene. This enzyme is not present in HTC cells. Conversely, TAT is inducible in the HTC cell by steroids and is not present or inducible in 3T3. The very low level of apparent TAT activity in 3T3 cells is not identical to the TAT in hepatoma cells. The 3T3 activity, unlike both mouse hepatic and HTC TAT, is not inactivated by antiserum monospecific to authentic TAT. Thus we could study the inducibility of two reciprocally inducible enzymes in the hybrids. We found that aryl hydrocarbon hydroxylase inducibility was greatly enhanced, but as in all the other cases that have been studied, the TAT inducibility was extinguished.

We have also fused HTC cells with the L cells of line L-B82 (3). The properties of the hybrids are summarized in Table 1. As usual, in the hybrids the inducibility of the TAT activity is extinguished. In these experiments (and also in the experiments with 3T3), we have mixed cell extracts, and the results showed that there are no obvious inhibitors of TAT activity in the hybrids. We have also examined the hybrids with monospecific antiserum against rat TAT and find no evidence for residual rat TAT in these cells. Since we did not find, as had Weiss, the reexpression of inducibility, the possibility remains that our hybrids have lost entirely the gene for rat TAT. However, one can make the usual arguments that, on the basis of total chromosome count and the fact that TAT extinction occurs in every hybrid, this interpretation seems unlikely. Since her fusions used a different inducible hepatoma than ours, the difference in our results may reside with an inherent difference in the cell lines.

Cyclic AMP in the HTC–L cell hybrids was measured by Al Gilman. The L cells have about 5 pmoles of cAMP per milligram protein, while the HTC

TABLE 1. *Summary of characteristics of HL hybrids and parental cells*

	L Cell (B82)	HTC (ARI)	HL (Hybrid)
Thymidine kinase	−	+	+
Hypoxanthine-guanine phosphoribosyl transferase	+	−	+
Total			
chromosomes	50	62	106 (95–120)
small metacentries	1–3	8–10	7–14
Tyrosine aminotransferase			
activity	low	high	low
immunoreactive	0	+	0
inducibility	0	+	0
Cyclic AMP (pmoles/mg)	5	1	5
C′2 complement production (molecules/hr/cell)	0	65	33
Catalase	low	high	high
Contact inhibition	−	−	+

cells have only 1 pmole of cAMP per milligram protein. The hybrids have the level present in the L cells.

The second component of complement (C′2) also was examined in the hybrids. C′2 production occurs in HTC cells but not in L cells. The hybrids were found to produce an intermediate level of C′2. Catalase activity, which is high in HTC cells, was also measured. This is not an ideal marker because there are low levels of the enzyme in L cells, and one must use quantitative, not qualitative differences. However, with these reservations, the hybrids had high activity, like the HTC parental cells. Therefore several biochemical markers corroborate the fact that these are hybrids and some show that not all specialized functions, such as C′2 production, are shut off.

The final phenotype that I will mention is what is listed in the table as "contact inhibition." Neither of the parental cell lines stopped growing at confluency. In contrast, the HTC–L cell hybrids did do so.

The parental and hybrid cells were morphologically quite distinct (Fig. 2) and their differing growth pattern obvious. As a biochemical verification of their behavior, we carried out labeling studies with ^{3}H thymidine (Fig. 3). When ^{3}H thymidine was added to a confluent colony of hybrid cells, the

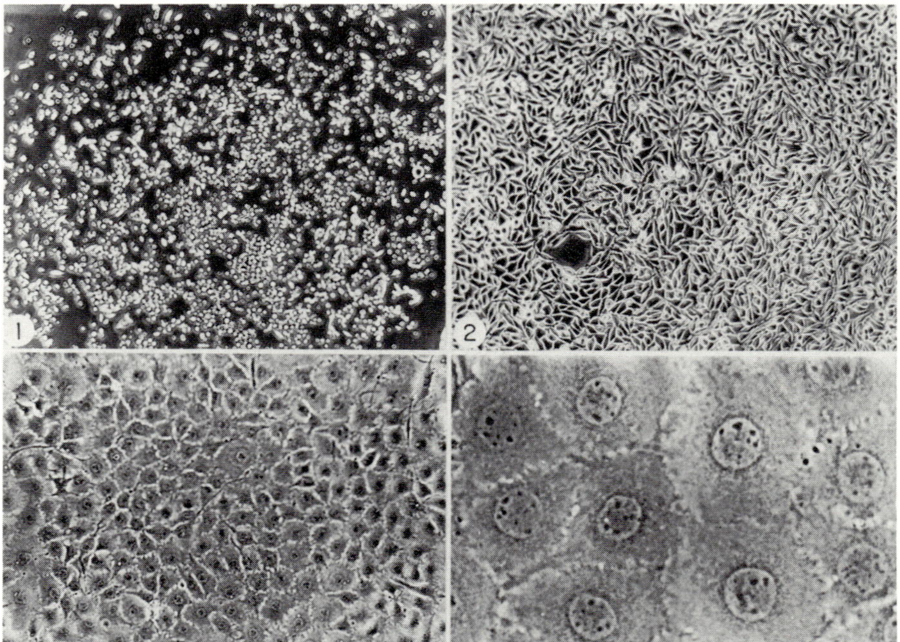

FIG. 2. Morphology of HTC cells (1), L-B82 (2) cells, and two different HTC–L-B82 hybrid clones (3 and 4). Phase contrast, Fields 1–3, × 125; 4, × 250. (From ref. 3.)

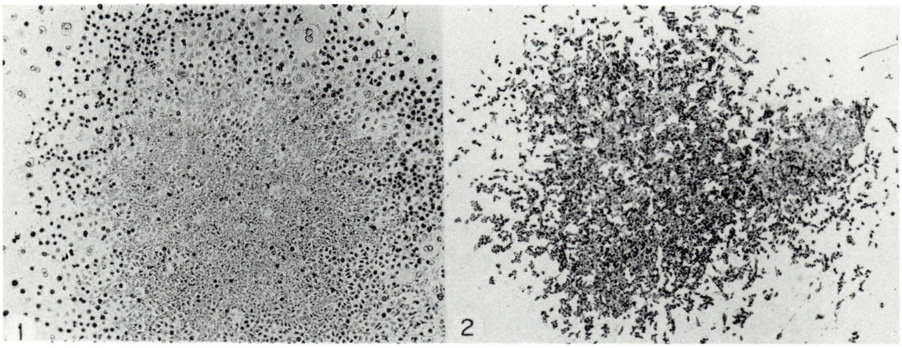

FIG. 3. ^{3}H thymidine labeling of confluent cultures of (1) HTC–LB82 hybrid clone HL5 (seen in field 3 of Fig. 2); (2) HTC cells. Bright field microscopy, ×40. (From ref. 3.)

labeled cells were predominantly found around the perimeter of the colony. In contrast, HTC cells become labeled throughout their colony. All the 20 clones we obtained in the original experiment were contact inhibited. We have now repeated the experiment several times, and each time we found some clones that behaved like the first hybrids, but also there were others which resembled an intermediate cell type morphologically and were not "contact inhibited." We are now analyzing these new hybrids biochemically and karyologically.

REFERENCES

1. Aviv, D., and Thompson, E. B. (1972): Variation in tyrosine aminotransferase in HTC cell clones. *Science 177*:1201–1203.
2. Benedict, W. F., Nebert, D. W., and Thompson, E. B. (1972): Expression of aryl hydrocarbon hydroxylase induction and suppression of tyrosine aminotransferase induction in somatic-cell hybrids. *Proc. Nat. Acad. Sci. USA 69*:2179–2183.
3. Levisohn, S. R., and Thompson, E. B. (1973): Contact inhibition and gene expression in HTC/L cell hybrid lines. *J. Cell. Physiol. 81*:225–230.
4. Weiss, M. C., and Chaplain, M. (1971): Expression of differentiated functions in hepatoma cell hybrids: Reappearance of tyrosine aminotransferase inducibility after the loss of chromosomes. *Proc. Nat. Acad. Sci. USA 68*:3026–3030.

DISCUSSION

Gerald: I am somewhat puzzled. I had thought that C2 production is a macrophage-specific function. Are there any other liver-specific functions expressed by these hepatoma cells?

Thompson: The cells exhibit the induction of TAT by steroid hormones, supposedly a very specific liver cell marker. We recently have been able to demonstrate that they excrete some serum proteins, as well. On the other hand, the hepatoma cells have no phenotype that I would associate with macrophages. Apparently, as tumor cells sometimes do, they are expressing a gene not normally expressed in their parental cell type. If you could suggest a good macrophage marker, we could test it.

Gerald: Gordon might suggest lysozyme.

Croce: Do your rat-mouse hybrids have the dexamethasone receptor?

Thompson: Yes, these hybrids possess normal levels of receptor activity. We have preliminary evidence that they have both the L-cell type and the HTC-type receptor, which can be distinguished. While the hybrids failed to express HTC functions in response to stimulation, they did express L cell functions.

Croce: Have you demonstrated the nuclear transfer of the dexamethasone receptor in the hybrids?

Thompson: No, we have not.

Somatic Cell Hybridization, edited by
Richard L. Davidson and Felix de la Cruz.
Raven Press, New York © 1974.

Multiple Forms of Tyrosine Aminotransferase in Hepatoma Cells: A Caution

Thomas D. Gelehrter

Departments of Human Genetics, Medicine, and Pediatrics, Yale University School of Medicine, New Haven, Connecticut 06510

I would like to discuss some work on tyrosine aminotransferase (TAT) in rat hepatoma (HTC) cells that, although not done with hybrids, is, I believe, relevant to studies of the regulation of TAT in hybrid cells. In order to determine whether there is separate genetic control of TAT expression and of its inducibility, it is important at the outset to know exactly what enzyme one is studying, i.e., do the basal and induced activities represent the same enzyme.

We have been investigating the mechanisms by which dexamethasone, insulin, and a macromolecular factor in serum cause an additive increase in the activity of TAT in HTC cells (1). Do these effectors induce isozymic forms of TAT, or do they affect different aspects of the control of a single protein? In collaboration with Janet Emanuel and Carolyn Spencer, I have analyzed extracts of HTC cells induced with dexamethasone alone, dexamethasone plus insulin, and dexamethasone plus serum by titration with specific antibody to purified rat liver TAT, by heat stability, and by electrophoresis on polyacrylamide gels. Our results indicate that the TAT induced by all three effectors is the same (2).

During the course of this study, however, we observed heterogeneity of TAT activity in HTC cells that had not previously been recognized (Fig. 1). If an extract of induced HTC cells is subjected to electrophoresis and stained histochemically for TAT, there is initially only a single band of activity apparent near the anode, designated H-1. With longer staining times two additional slowly migrating bands appear, designated H-2 and H-3. Only the more anodal form, H-1, appears to increase in amount after incubation of HTC cells with dexamethasone, insulin, or serum.

In all cases where one sees multiple forms of an enzyme, it must be determined whether they are simply aggregates of a single protein or whether they might represent isozymes. Figure 2 shows a Ferguson plot, in which the relative mobility of the three forms of TAT is plotted against the concentration of polyacrylamide in the gels. The slopes of the major band of hepatoma cell TAT activity (H-1) and the slowest band (H-3) are parallel, which indicates that these two bands have the same retardation coefficient. Since

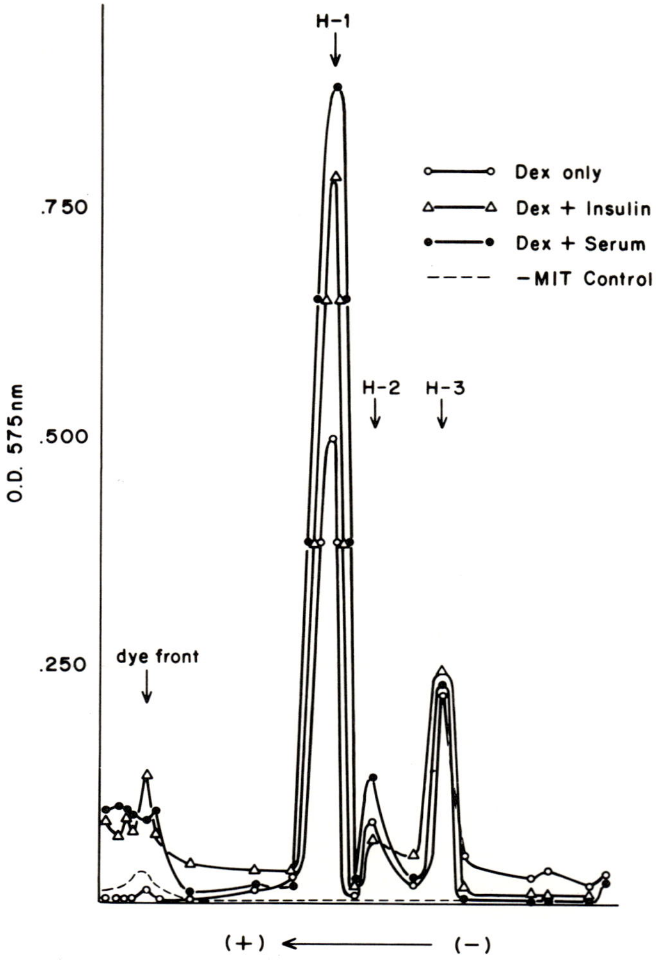

FIG. 1. Polyacrylamide gel electrophoresis of HTC cell tyrosine aminotransferase induced by dexamethasone alone, dexamethasone plus insulin, and dexamethasone plus serum. HTC cells in suspension culture were induced with dexamethasone alone, dexamethasone plus insulin, or dexamethasone plus serum. Cytoplasmic extracts were prepared and heated at 65°C for 6 min to achieve a threefold purification. 200 μl portions of these extracts, containing 7 to 26 mU of TAT activity, were loaded on 7.5% gels; electrophoresis was performed for 130 min at room temperature at 3.5 mA per gel. Histochemical assay for TAT was performed at 60°C for 75 min. The stained gels were scanned at 575 nm, and the tracings superimposed for comparison. The stained bands are arbitrarily designated H-1, H-2, H-3, according to their mobility toward the anode. An additional gel was stained without the substrate monoiodotyrosine (MIT) to demonstrate the specificity of the staining method (— — —). (From ref. 2.)

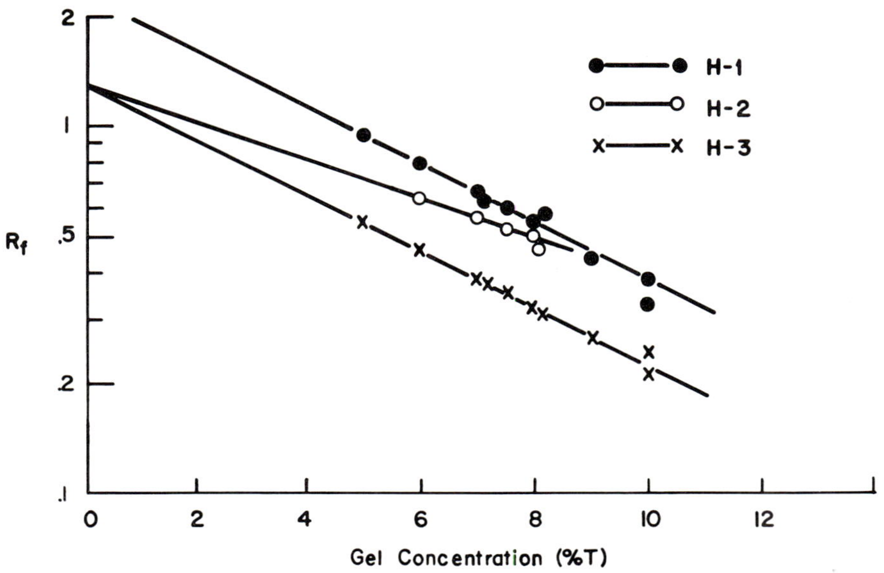

FIG. 2. Polyacrylamide gel electrophoresis of HTC cell tyrosine aminotransferase at different gel concentrations. Data from two separate experiments are shown. In both, suspension cultures of HTC cells were induced with dexamethasone, and heated, cytoplasmic extracts prepared as in Fig. 1. Approximately 15 mU of TAT activity was loaded on each gel. Electrophoresis was performed at 3 mA per gel for 140 or 210 min. Gel concentration ranged from 5 to 10%. Staining was performed at 60° for 65 or 85 min. The relative mobility (relative to the dye front) is plotted on a log scale against the gel concentration. H-1, H-2, H-3 refer to the same three bands described in Figure 1. (From ref. 2.)

the retardation coefficient is related to molecular weight, this suggests that these two forms of TAT have the same molecular weight but differ in molecular charge. Hence they might represent isozymes. The middle band of TAT activity (H-2), which was not always seen, had the same Y-intercept as the slower band (H-3) indicating that these two forms have the same valence or net charge but differ in molecular weight. The simplest explanation is that H-2 is an enzymatically active oligomer of H-3 (2).

What is the significance in the hepatoma cells of these two forms of TAT that differ in molecular charge? A mitochondrial TAT isozyme has been described; but this enzyme is now thought to be identical to mitochondrial aspartate aminotransferase (4). The slow band of TAT activity is not this mitochondrial transaminase. First, it is heat stable, whereas the mitochondrial enzyme is heat labile; second, both H-1 and H-3 are found in the cytoplasm of HTC cells and not in the mitochondria (6).

Ruddle raised the possibility that the second band might be glutamate-oxaloacetate transaminase (GOT), or more correctly aspartate aminotransferase. When parallel gels are stained for GOT activity, we find a band of GOT activity in the position of the H-3 enzyme. If gels of different acrylamide concentrations are used, it is not possible to separate the GOT activity from the minor band of TAT activity, suggesting that both activities are the result of a single enzyme (6). According to the literature, however, purified soluble rat liver TAT does not transaminate aspartate (3), and purified soluble GOT does not transaminate tyrosine (4). The H-3 enzyme appears to transaminate both tyrosine and aspartate.

The H-3 form of TAT also differs from the usual hepatocyte TAT in that it can utilize oxaloacetate in place of α-ketoglutarate as the ketoacid in the transamination of tyrosine, whereas the H-1 enzyme is specific for α-ketoglutarate. Furthermore, if the TAT histochemical assay is carried out in the presence of L-aspartate, the H-3 enzyme is inhibited, whereas the H-1 enzyme is not, again indicating a difference in their catalytic activities (6).

Finally, if one incubates an extract of HTC cells with antiserum to purified soluble rat liver TAT, the major form of TAT (H-1) is inactivated while the H-3 form is unaffected. Conversely, if one incubates the extract with antibody to purified soluble pig heart GOT (kindly provided by Martinez-Carrion of Notre Dame University), the H-3 form is inactivated, whereas the H-1 enzyme is unaffected (6). Our results indicate, therefore, that the H-3 enzyme is a "pseudoisozyme" of TAT. We initially thought we had multiple forms of TAT; in fact, there is another enzyme in the cell, aspartate aminotransferase, which, because of its broad substrate specificity, is capable of transaminating tyrosine (5).

What is the relevance of these observations to studies in hybrid cells? First, many studies in somatic cell genetics are based on analyses of isozymic patterns on starch gels. Clearly it is very important to have good biochemical and genetic characterization of such isozymes in order to know precisely which enzyme one is studying. Second, our observations may also be relevant to studies on the reappearance of TAT activity in hybrids, especially where one is dealing with very low activities of the enzyme (7). If the hybrids have relatively high levels of GOT, which as we have shown has the ability to transaminate tyrosine, then the "turned on" activity in the segregant hybrids could be the result of this enzyme. This possibility is particularly important because the induction kinetics of TAT in the parent hepatoma cell and in the hybrids which reexpress the function have been reported to differ (7). Thus one must be very cautious about interpreting the mechanism of apparent reexpression of TAT activity.

Ideally one would like to produce hybrids in which one parent cell has inducible TAT and the other parent has noninducible but easily measurable basal TAT activity, and in which the two parental forms of TAT could be differentiated by electrophoretic mobility and other characteristics. It might then be possible to obtain less equivocal information about the genetic regulation of TAT induction.

REFERENCES

1. Gelehrter, T. D., and Tomkins, G. M. (1970): Posttranscriptional control of tyrosine aminotransferase synthesis by insulin. *Proc. Nat. Acad. Sci. USA 66*:390–397.
2. Gelehrter, T. D., Emanuel, J. R., and Spencer, C. J. (1972): Induction of tyrosine aminotransferase by dexamethasone, insulin, and serum: Characterization of the induced enzyme. *J. Biol. Chem. 247*:6197–6203.
3. Kenney, F. T. (1959): Properties of partially purified tyrosine-α-ketoglutarate transaminase from rat liver. *J. Biol. Chem. 234*:2707–2712.
4. Miller, J. E., and Litwack, G. (1971): Purification, properties, and identity of liver mitochondrial tyrosine aminotransferase. *J. Biol. Chem. 246*:3234–3240.
5. Shrawder, E., and Martinez-Carrion, M. (1972): Evidence of phenylalanine transaminase activity in the isoenzymes of aspartate transaminase. *J. Biol. Chem. 247*:2486–2492.
6. Spencer, C. J., and Gelehrter, T. D. (1974): Pseudoisozymes of hepatic tyrosine aminotransferase. *J. Biol. Chem. 249*:577–583.
7. Weiss, M. C., and Chaplain, M. (1971): Expression of differentiated functions in hepatoma cell hybrids: Reappearance of tyrosine aminotransferase inducibility after the loss of chromosomes. *Proc. Nat. Acad. Sci. USA 68*:3026–3030.

DISCUSSION

Krooth: Is the second enzyme, the pseudoisozyme, tissue specific? Is it found only in liver or is it found in a variety of tissues?

Gelehrter: The GOT is apparently not tissue specific. The cytoplasmic GOT is found in a variety of tissues, including the kidney. The antibody to cytoplasmic GOT is not tissue or species specific but is specific for the cytoplasmic as opposed to the mitochondrial form of the enzyme. The antibody to TAT is also not species specific. The major band of mouse and rat liver TAT and the TAT of the hepatoma cells are all inactivated by the antibody. The enzyme, however, is apparently tissue specific.

Somatic Cell Hybridization, edited by
Richard L. Davidson and Felix de la Cruz.
Raven Press, New York © 1974.

Genetic Analysis of the Nervous System Using Somatic Cell Hybrids

John Minna and Alfred Gilman

Laboratory of Biochemical Genetics, National Heart and Lung Institute, National Institutes of Health, Bethesda, Maryland 20014 and Department of Pharmacology, University of Virginia, Charlottesville, Virginia 22903

I would like to describe two pieces of work on the genetic analysis of the nervous system. The first area deals with the regulation of intracellular adenosine 3′5′ cyclic monophosphate (cyclic AMP) in hybrid cells. The experiments were carried using a sensitive protein binding assay for cyclic AMP described by Gilman. A complete description of these experiments is published elsewhere (1, 2).

The central dogma of cyclic AMP metabolism states that the cellular specificity of response to hormones that regulate intracellular levels of cyclic AMP resides in the presence of specific receptors in the cell membrane. Then, through an unknown series and number of events, adenyl cyclase activity is stimulated and cyclic AMP is synthesized.

We have been studying the effects of exogenously applied catecholamines and prostaglandin E_1 (which will be referred to below as PGE_1) on a variety of parental and hybrid cells.

The catecholamine response for hybrids of rat glial cells and mouse fibroblasts is shown in Table 1. The glial cells show a 300-fold increase in intracellular cyclic AMP (from 10 to 3,000 pmoles per milligram of cell protein) after catecholamine stimulation and were scored as β^+. In contrast, the fibroblasts showed no effect of catecholamine stimulation and were scored as β^-.

A series of hybrid clones showed less than 1% of the response of the β^+ parent when exposed to catecholamine, and the hybrids thus were scored as β^-. Chromosome analysis showed no evidence for loss of the rat glial cell marker chromosomes at the time the hybrid cells were studied. Thus, the results suggested that β^+ cells fused with β^- cells yield β^- hybrids.

When tested with PGE_1, clonal differences in cyclic AMP accumulation were found. For example, B82 cells showed a three- to four-fold increase in cyclic AMP after PGE_1 treatment, while the mouse renal adenocarcinomal, RAG, showed no response. A series of hybrid clones were all responsive to PGE_1, with levels of response equal to or greater than PGE_1^+ parent (Table

TABLE 1. *Effect of Catecholamine on C6TG1A-B82 hybrid cells*

	Cyclic AMP pmoles/mg protein		Mean chromosome number markers[b]		
Cell line[a]	Control	Norepinephrine (0.1 mM)	Total	C6TG1A	B82
Parent					
C6TG1A	10	3000	42	20	0
B82	14	13	52	0	26
			94	20	26
Sample Hybrids					
CB4	18	36	92	22	28
CB7	7	12	93	22	22
CB pool (100 clones)	16	23	93	22	26

[a] None of 12 independently derived hybrid clones, or two pools of 100 hybrid colonies each, showed a large response to exogenously applied norepinephrine or isoproterenol. Sample hybrid clones shown represent extreme values found.

[b] C6TG1A markers are acrocentric and small metacentric chromosomes while B82 markers are large metacentric and large submetacentric chromosomes.

2). The conclusion of these series of experiments is that PGE_1^+ cells fused with PGE_1^- cells yield PGE_1^+ hybrids.

From the results of a number of crosses, we can say that, when cells that show a very large quantitative difference in the β^+ response are fused together, the hybrids that result from these matings exhibit the same or lower levels of response than the parental cell with the lower response. In contrast, when cells that show very large quantitative differences in the PGE_1 response are fused together, the resulting hybrids are at least as responsive to PGE_1 as the parental cell with the higher response, and sometimes more responsive.

TABLE 2. *Effect of PGE_1 on RAG-B82 Hybrid clones*

	Cyclic AMP pmoles/mg protein		Mean chromosome number	
Cell line[a]	Control	PGE_1 (2 μg/ml)	Total	Large metacentric
Parent				
RAG	5	5	67	5
B82	14	48	52	26
			119	31
Sample Hybrids				
RB4B	31	110	82	23
RB2A	12	57	108	31

[a] Sample hybrids represent extreme values found in the hybrid clones tested. The hybrid nature was confirmed by glucose phosphate isomerase isozyme analysis. RB hybrids were GPI-1AB.

In addition to the above responses, it is possible to study the potentiation of peak hormone response by the phosphodiesterase inhibitor theophylline. Theophylline plus and minus cells were fused together, and the hybrids were found to be theophylline plus (data not shown). Thus, it is possible to use the response to theophylline in terms of intracellular cyclic AMP concentration following hormonal stimulation as a genetic marker. Since theophylline is a phosphodiesterase inhibitor, these inherited differences of theophylline sensitivity presumably reflect differences in cyclic nucleotide phosphodiesterase.

I would now like to describe briefly the attempts Coon and I have made to try to get normal nervous tissue to replicate *in vitro*. Our plan of attack was to fuse transformed human fibroblasts (VA-2) with fresh trypsinized suspensions of cells dissected out from specific regions of the rodent embryonic nervous system. Based on the results of other systems, we expected that the human fibroblast should be able to initiate DNA synthesis and cell division in the mitotically quiescent rodent cells of the nervous system. We also anticipated that the human chromosomes would be lost from the hybrid cells, and that it might then be possible to isolate hybrids expressing neural functions.

We have isolated propagating hybrid cell lines from a variety of species and tissues of adult and embryonic sources (Table 3) and have studied their properties. The hybrids can be isolated without difficulty and with good yields. However, there may be differences in the frequency of hybridization achieved with embryonic nervous system cells and with multiplying fibroblast lines.

In order to screen many colonies rapidly for neural functions, we used a specific histochemical stain for acetylcholinesterase (3). In colonies that have activity, a dark red-brown precipitate forms, while clones with no activity show no precipitate formation. The assay is capable of detecting levels

TABLE 3. *Normal tissues which have yielded propagating hybrid cell lines when fused to human fibroblasts (VA-2)*

Species	Tissue
Mouse	
Embryonic	Brain, spinal cord, neural retina, dorsal root ganglia, liver, and heart
Adult	Bone marrow, spleen, thymus
Rat	
Embryonic	Brain, spinal cord, neural retina, dorsal root ganglia, superior cervical ganglia, heart
Chinese hamster	
Embryonic	Brain, neural retina, spinal cord, heart, liver
Adult	Bone marrow, spleen

of 10 to 20 nmoles of product formed per milligram of protein per minute. The level with normal neurons is at least one order of magnitude higher than this. All our attempts thus far have been unsuccessful. Approximately 4,000 colonies arising in fusions between fibroblasts and normal embryonic nervous tissue cells have been tested, and no definite acetylcholinesterase positive colonies have been found.

It is of interest that all of the hybrid colonies (over 200) arising in crosses between mouse neuroblastoma cells and Chinese hamster brain cells were acetylcholinesterase positive. I think that it would be interesting to examine the results of fusion of neuroblastoma cells with cells of other tissues and also the results of fusing other cell lines expressing differentiated functions with the normal cells of their tissues of origin.

Part of the explanation for the lack of expression of acetylcholine in hybrids between human fibroblast and normal mouse nervous cells may come from the finding that greatly surprised us—namely, that in the vast majority of the hybrids, human chromosomes were retained while mouse chromosomes were being preferentially lost. Coon has discussed this aspect.

Additional evidence for loss of mouse genes was obtained by isozyme analysis. Some examples are shown in Table 4. Very few clones express all the mouse isozymes. The majority have lost some, or all, of the mouse gene products. I should stress that to date the hybrids that do not express all the mouse isozymes always express all the human isozymes, a finding in great contrast to that usually found in mouse-human hybrids.

TABLE 4. *Examples of isozyme phenotypes in human fibroblast (Va2)-mouse nervous system hybrids*

	Mouse isozyme present								
Cell line	GPI	PGM2	LDHA	IDH	G6PD	IPO	GOT	PEPA	PEPB
VMSG2	+	+	+	+	+	+	+	+	+
VME15	+	−	+	−	−	−	+	−	+
VMPTH28G	−	−	−	−	+	−	−	−	−
VMSG4	−	−	−	−	−	−	−	−	−

In conclusion, while our search for replicating neural cells thus far has been unsuccessful, these hybrid lines appear to present a new type of chromosome segregation. They can be used to map mouse genes and also as a source of cells that retain certain human chromosomes that have very infrequently or never been found in hybrid cells. The hybrids can also be used to study patterns of chromosome segregation in hybrids derived from normal tissue and should provide material of unique cellular composition for other cell biology studies.

REFERENCES

1. Gilman, A., and Minna, J. (1973): Expression of genes for metabolism of adenosine 3′,5′-cyclic monophosphate in somatic cells. I. Responses to catecholamines in parental and hybrid cells. *J. Biol. Chem.* *248*:6610.
2. Minna, J., and Gilman, A. (1973): Expression of genes for metabolism of adenosine 3′,5′-monophosphate in somatic cells. II. Effect of prostaglandin E_1 and theophylline on parental and hybrid cells. *J. Biol. Chem.* *248*:6618.
3. Minna, J., Glazer, D., and Nirenberg, M. (1972): Genetic dissection of neural properties using somatic cell hybrids. *Nature New Biol.* *235*:225.

DISCUSSION

Koprowski: I do not understand which type of nervous cells you used for hybridization. You mentioned spinal cord and brain. Have you separated the cells?

Minna: I have used the terms "nervous tissue" or "tissue from the nervous system." Obviously, if one takes sections of whole brain even in early embryonic development, only a fraction of the cells would be neurons. We generally use 13- to 18-day-old embryos and dissect out various regions of the brain. We have been trying to use methods to purify neurons out of these primary suspensions. The easiest and gentlest methods involve attachment to a substrate, since normal neurons do not attach very rapidly. Another method might involve some antimetabolite that would kill off the fibroblasts.

Shows: In the last table, were the results for hybrids between VA-2 and mouse cells, and were the hybrids grown in HAT medium? There was one clone with mouse G6PD, but the other clones were negative for mouse G6PD.

Minna: One of the hybrids did have mouse G6PD. Every time that we have tested hybrids maintained in HAT medium, both human and mouse G6PD are retained. If the HAT selection is removed, the hybrids very rapidly lose the mouse G6PD. Such hybrids subsequently will not grow in HAT. Since the human line, VA-2, is deficient in HGPRT, hybrid lines that grow in HAT medium presumably should retain the mouse gene for HGPRT. Although we have not yet specifically assayed for HGPRT by electrophoretic means, one simple assumption would be that hybrids that can't grow in HAT have lost the mouse X chromosome.

Shows: Were there specific human linkage groups retained?

Minna: These hybrids are perfect for studying mouse rather than human linkage, since they lose mouse chromosomes. We are just putting all the data together on this point. We hope that others will become interested in using these types of hybrids, because they may give a new input into the study of linkages.

Shows: How many human chromosomes are retained in the hybrids?

Minna: It appears that a full human set is usually retained. In the majority of the hybrids, the human isozymes of all the enzymes studied were present whereas mouse isozymes are lost. Human isozymes have been lost from a few of the hybrids, but this is the exception rather than the rule.

Coon: I could point out that, although we had frequently seen the loss of all of the rodent chromosomes from the hybrids, we have never seen the loss of all of the human chromosomes. It looks almost as though it may be necessary to retain a few human chromosomes in these hybrids in order to keep them multiplying.

Minna: Since we are obviously applying continual pressure to select for cell growth, theoretically one should be able to identify some human chromosomes from the SV 40 transformed parent that play a role in maintaining continued growth. Another question that we could ask is whether a linkage group of one species could function to completely replace the corresponding linkage group of the other species. I do not think that that has been seen yet.

Pontecorvo: On first principles, it is most unlikely that you will ever find it.

Minna: We will just have to wait for the data.

Somatic Cell Hybridization, edited by
Richard L. Davidson and Felix de la Cruz.
Raven Press, New York © 1974.

Deficiency of the Fourth Component of Complement (C4): Studies on the Molecular Basis of the Genetic Abnormality*

Harvey R. Colten†

Allergy Division, Department of Medicine, Children's Hospital Medical Center, and the Department of Pediatrics, Harvard Medical School, 300 Longwood Avenue, Boston, Massachusetts 02115

I will discuss *in vitro* biosynthesis of serum proteins and the use of hybrid cells in an analysis of some of the genetic factors that regulate serum protein production. In particular, I will discuss the serum complement system. Complement consists of nine serum proteins that interact to mediate most of the features of the inflammatory response. In these studies we used cells derived from a strain of guinea pigs genetically deficient in the fourth component of complement (C4). The serum of homozygous deficient animals contains no detectable C4 even though the methods used for detection were capable of measuring as little as one one-millionth the normal concentration.

Using methods developed in our earlier studies of complement biosynthesis, we examined the capacity of cells isolated from C4-deficient animals to synthesize C4 *in vitro*. In these and most of the other studies to be described, C4 was measured with an extremely sensitive and specific functional assay, capable of measuring as few as 5×10^6 to 1×10^7 effective molecules of C4 per ml. Because of the specificity of this functional assay, it is not necessary to isolate the C4 from complex mixtures such as tissue culture media. In these experiments (performed with Michael Frank), we found that cells from homozygous deficient animals produced no detectable C4 in culture. In contrast, cells from normal animals produced substantial amounts of C4 (2). Cells from heterozygous deficient animals produced C4 at a rate intermediate between that of the normal and the homozygous deficient animals.

In an attempt to determine the basis for the C4 deficiency, Robertson Parkman and I fused peritoneal exudate cells (the cells that in the normal animal produce C4) from C4-deficient animals with HeLa cells. The hybrid cells were cloned and subcloned, and the parental cells and the hybrids examined for their capacity to produce C2 (the second component of com-

* Supported by USPHS grants HD-05916 and AI-05877.
† Recipient of USPHS Career Development Award 1-K4-HD-GM-70,558.

plement) and C4. As mentioned above, C4-deficient cells produced no detectable C4 *in vitro* whereas they were capable of producing guinea pig C2. HeLa cells produced no detectable C2 or C4. However, several but not all of the hybrid clones produced C4. The C4 produced by the hybrids was shown to be human in origin on the basis of its functional characteristics and also on the basis of its immunochemical identity with human C4. The hybrids produced neither guinea pig nor human C2. These results will not be presented in detail because they have been published (3).

Following these observations, we performed additional experiments designed to explore the molecular basis of this surprising result. The results were particularly unexpected since most of the previous experiments with hybrids had suggested that specialized functions are generally extinguished and that no new functions appear. We examined a soluble extract of C4-deficient cells and also medium harvested from cultures of C4-deficient cells. The cell-free media or extracts were added to growing cultures of HeLa cells. At timed intervals after addition of this material derived from the C4-deficient cells, aliquots of the media were removed from the HeLa cell culture and examined for the appearance of human C4 activity.

In cultures exposed to the material derived from the C4-deficient cells, after a lag period, there was a marked production of human C4 (Fig. 1).

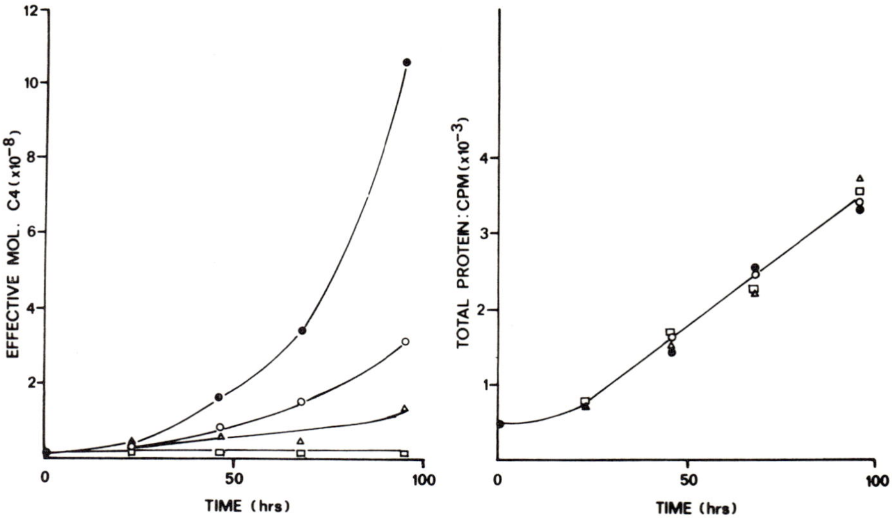

FIG. 1. Effect of regulator on C4 synthesis and on total protein synthesized and secreted into the medium: HeLa cells were incubated in medium harvested at 48 hr from (●) $C4D_1$ (C4-deficient cells) (1×10^6 cells, (○) from $C4D_1$ diluted 1:5 with fresh medium 199, (△) from HVW (normal guinea pig cells) (1×10^7 cells), and in fresh medium 199 (□). (A) human C4; (B) total protein. (From ref. 1.)

Control populations exposed to a medium alone or to a medium derived from a variety of other cells produced no detectable C4.

Simultaneous measurement of incorporation of ^{14}C-labeled amino acids into TCA-precipitable material indicated that the factor derived from the C4-deficient cells had no effect on total protein synthesis in the HeLa cell. Thus, there was no generalized stimulation of the HeLa cell by this factor (1). Production of the factor by C4-deficient cells and the response to it by HeLa cells were both inhibitable by relatively low concentrations of actinomycin D (1).

The factor from the C4-deficient cells was found to be heat stable (at 56°C for 1 hr), nondialyzable, and partially inactivated by RNAase and trypsin but not by DNAase. Chromatography on Sephadex G-100 demonstrated that the activity eluted in a position corresponding to a 45,000 molecular weight marker (Fig. 2).

In our initial experiments, we noted a substantial lag between the time of addition of the factor and the onset of detectable C4 production by the HeLa cells. This observation suggested the possibility that the response to the factor might be a function of the cell cycle. We tested this possibility with synchronized HeLa cells (synchronized by shaking them from actively growing cultures). At timed intervals after synchronization, medium containing the factor was added to the HeLa cell preparations. The results indicated that the duration of this lag period was a function of the time dur-

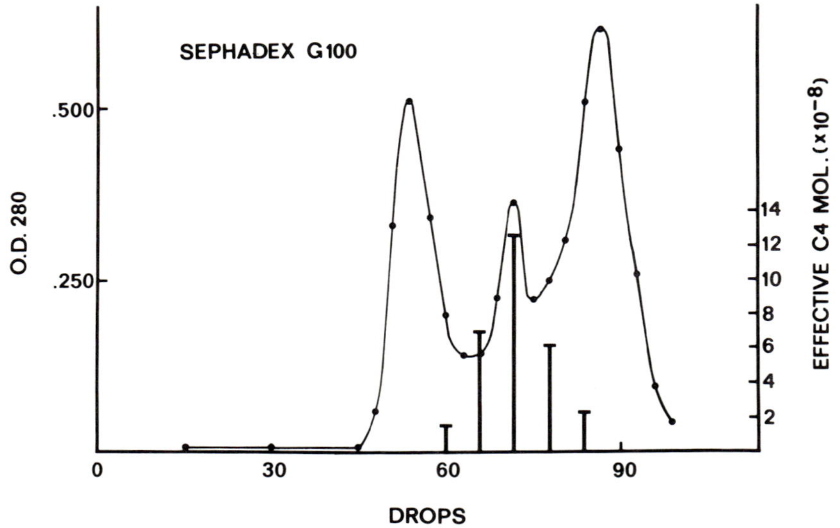

FIG. 2. Chromatography of C4-deficient regulator factor. Sephadex G100 column calibrated with 0.5 cc normal guinea pig serum (OD 280●−−−−●). Regulator activity indicated by solid bars, assayed on HeLa cell cultures.

ing the cell cycle at which the factor was added to the cells (see Fig. 3). For example, if the factor was added shortly after synchronization, i.e., while the cells were in G-1, there was a lag period of approximately 17 to 18 hr before the appearance of detectable C4 production. On the other hand, if the factor was added at approximately the time of onset of DNA synthesis (as detected by the incorporation of tritiated thymidine), the lag period was somewhat less than 8 hr. Additional experiments with short exposures to the factor indicated that the HeLa cells were responsive to the factor in early *S* phase (4). A number of possible interpretations for this observation have been considered and are currently under investigation.

The studies outlined above were initiated in an attempt to learn something about the genetic abnormality in production of complement in C4-deficient guinea pigs. I think they should be interpreted with some caution inasmuch as we may be learning more about the HeLa cells than the C4-deficient guinea pig. I will mention one recent finding that leads me to that conclusion. Some but not all fetal sera are also capable of inducing C4 production in HeLa cells. We now have evidence that the factor in fetal serum, whatever it may be, is quite distinct from the factor produced by the C4-

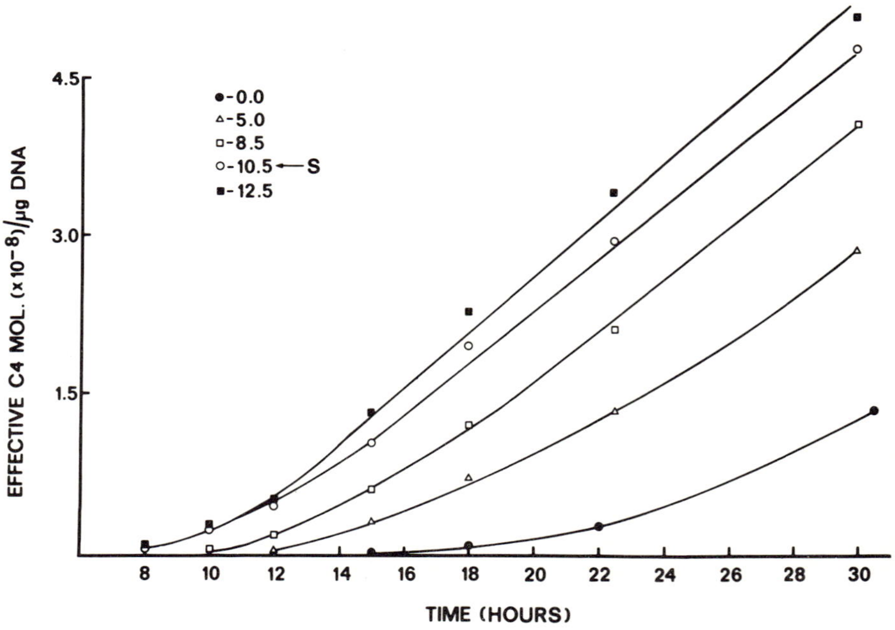

FIG. 3. Effect of cell cycle on response of HeLa cells to the regulator factor obtained from C4-deficient cells. Medium containing the factor added at times (hours) after synchronization of HeLa. *S* Phase indicated by arrow. Abscissa indicates time of incubation of HeLa in the presence of the factor.

deficient cells. First, not all HeLa cells respond to the serum factor, but thus far all the HeLa cells that we have examined do respond to the C4-deficient factor. Second, when the serum is subjected to Sephadex chromatography, there is a diffuse elution pattern with no clear-cut peak of activity in the 45,-000 molecular weight region. Third, normal guinea pig peritoneal cells are responsive to the serum factor but not responsive to the C4-deficient factor. Thus, the stimulating factors in serum and C4-deficient cells can be separated. The final interpretation of the original experiments, however, must await further work since it appears that induction of complement synthesis in HeLa cells may be accomplished by several different factors.

REFERENCES

1. Colten, H. R. (1972): *In vitro* synthesis of a regulator of mammalian gene expression. *Proc. Nat. Acad. Sci.* *69*:2233–2236.
2. Colten, H. R., and Frank, M. M. (1972): Biosynthesis of the second and fourth components of complement *in vitro* by tissues isolated from guinea pigs with genetically determined C4 deficiency. *Immunology* *22*:991–999.
3. Colten, H. R., and Parkman, R. (1972): Biosynthesis of the fourth component of complement (C4) by C4-deficient guinea pig-HeLa cell hybrids. *Science* *176*:1029–1031.
4. Stacey, D., and Colten, H. R.: *Unpublished observations.*

DISCUSSION

Davidson: Does the factor from the C4-deficient cells induce C4 production by any cells besides HeLa?

Colten: We have examined several cell types and thus far have not found any other cell, including C4 deficient or normal guinea pig fibroblasts, that responds.

Basilico: Is the induction restricted to the C4 component?

Colten: We have looked at other components of complement and have not found any others being produced. We have not yet looked at production of any other serum proteins.

Gerald: Is it possible that what you are calling human C4 is not actually C4 but is an analogous molecule?

Colten: The material produced by the HeLa cell in response to the guinea pig factor is immunochemically identical with human C4. On Ouchterlony analysis, there is a continuous line of precipitation between normal human C4 and the material produced by the stimulated HeLa cells, i.e., there is a line of complete identity (see Discussion Fig. 1). These results, along with the results of the functional assays, indicate that this is human C4 both functionally and immunochemically.

Gordon: Do normal guinea pig macrophages produce a similar factor and is it the same as the factor produced by the deficient cells?

Colten: Normal guinea pig cells produce the factor in extremely small amounts. There is, thus, a quantitative difference between the normal and

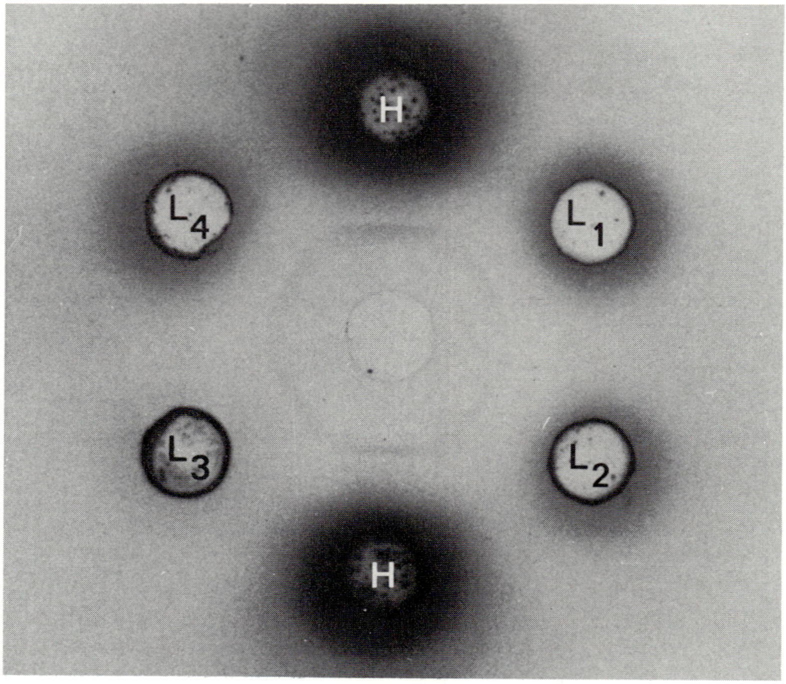

FIG. 1. Radioautograph of C4 produced by stimulated HeLa cells: center well contains rabbit anti-human C4. L_1, L_2, L_3, L_4—medium harvested from different normal human liver cultures (tissues incubated in medium containing ^{14}C-labeled amino acids). H—medium harvested from HeLa cultures grown in medium containing the regulator factor and ^{14}C-labeled amino acids.

the C4-deficient guinea pig cells with respect to production of the factor (see ref. 1). We have not yet analyzed the factor produced by the normal cells to see whether it is similar to the serum factor or to the C4-deficient factor.

Somatic Cell Hybridization, edited by
Richard L. Davidson and Felix de la Cruz.
Raven Press, New York © 1974.

Experiments Relating to the Use of Somatic Cell Hybrids for Studying the Control of Tissue-Specific Function

Robert S. Krooth

Department of Human Genetics and Development, College of Physicians and Surgeons, Columbia University, 630 West 168 Street, New York, New York 10032

My laboratory is not primarily devoted to somatic cell hybridization, and therefore my comments will deal partly with matters that are slightly peripheral to that subject.

Figure 1 shows the pathway for the catabolism of purines. The first two reactions in this sequence are catalyzed by the enzyme xanthine oxidase. In man, xanthine oxidase has a narrow and quite sharp tissue distribution. It is found in liver, gastrointestinal epithelium, and, as an extracellular enzyme, in milk. It is not demonstrable, even with extremely sensitive assays, in other tissues and it is absent from cultured human fibroblasts. In man and certain other primates, the catabolism of purines stops at uric acid. However, other mammals, such as rodents, are able to carry the purine catabolic pathway one step further, for these species have the enzyme uricase which catalyzes the conversion of uric acid to allantoin.

We have recently found that in rodents xanthine oxidase activity is not narrowly distributed but is found in most tissues and in most lines of cultured rodent cells (1). All the human lines tested, both homonuclear and heteronuclear, have no demonstrable xanthine oxidase, whereas most rodent lines do.

Interestingly enough, preliminary experiments suggest that in rodents uricase, which in that species is the final enzyme of the catabolic pathway, is limited to liver. In other words, it would appear that in the two species—man and rat—it is always the final enzyme that is tissue restricted.

What has all of this to do with somatic cell hybridization? I think there are two points of relevance. First, it would appear that an enzyme that is tissue specific in one species is not necessarily tissue specific in another. Xanthine oxidase is one example, but I strongly suspect that there are others. Hence, in interspecific somatic cell crosses involving the analysis of tissue specific function, it is important to verify that the function is indeed tissue specific in *both* parental species. In the case of some functions, such as the synthesis of albumin, I would be astonished if there were differences in

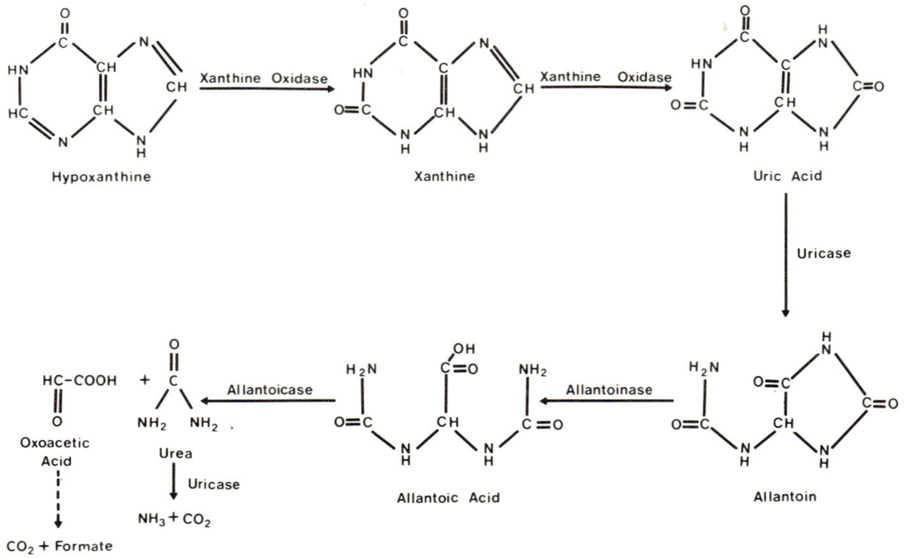

FIG. 1. Pathway for the catabolism of purines.

tissue distribution within the mammalian class. However, in the case of enzymes, such as tyrosine aminotransferase, species differences may exist and may lead in some cases to an erroneous interpretation of the results of somatic cell crosses.

My second point is a related one but slightly more interesting. I think that it would be worth inquiring, using existing hybrids formed between human and rodent cells, whether the hybrids contain xanthine oxidase activity. In most of the crosses thus far reported, where the behavior of a tissue-specific function was studied, the function involved was presumably tissue specific in *both* parental species. In the case of xanthine oxidase, one would be dealing with a function that is tissue specific in one parental species but not in the other. The rules that apply to such functions may be different from those that apply in cases where the tissue specificity of the function is demonstrable in both parental species.

I would now like to discuss one particular cross that our laboratory has been attempting to perform for the past 3 months. The experiment involves the hybridization of an established line of mouse erythrocytoid (F-2) cells (2) that make hemoglobin with a line of human homonuclear fibroblasts from a patient with the Lesch-Nyhan syndrome. The erythrocytoid cells were obtained from Charlotte Friend. One of the goals of the experiment was to see whether the hybrid would synthesize hemoglobin and of what type. The initial experimental design which we employed was very simple.

The erythrocytoid line will grow in suspension culture but not in monolayer. The Lesch-Nyhan fibroblasts will grow only in monolayer, but will not grow in HAT medium. The experiment then is to mix the cells together in suspension, in the presence of UV-inactivated Sendai virus, and subsequently to plate them in HAT medium and look for colonies growing in monolayer. Although this system works well for the isolation of hybrids between human leukocytes and mouse fibroblasts, in the present cross we have not thus far been able to isolate hybrids.

I should like briefly to mention what seem to be the difficulties and the way we now propose to surmount them. Possibly because the F-2 cells are so specialized in the erythrocytic direction, even low titers of inactivated Sendai virus destroy their ability to grow (and also to retain vital stains). Of course the killing of the F-2 cells by virus need not interfere with their ability to fuse with the fibroblasts. However it might select against any hybrid cells that resembled erythrocytes. The same may be true of the requirement that the hybrid be able to grow in monolayer. Another problem was that, in the absence of virus, it was remarkably difficult to wash away the erythrocytoid cells growing in suspension above the fibroblasts. This difficulty arose only in the absence of virus, because the virus kills the erythrocytoid cells. In the absence of virus, the erythrocytoid line persisted even when the cultures had been repeatedly washed on successive days with cell-free medium. Following every wash, a few erythrocytoid cells remained, and these eventually grew back to the point where they were present in overwhelming numbers. It now appears that the erythrocytoid lines are able to grow from exceedingly low cell population densities – a surprising result since nearly all lines that can proliferate in suspension will not grow in large volumes of liquid medium, unless they are at a high initial population density.

To get around one of these difficulties, we have now obtained, through the generosity of W. Ostertag and his associates, a line of erythrocytoid cells (3) that are deficient in thymidine kinase and will not grow in HAT medium. This should solve the problem of purging the culture of erythrocytoid cells following each opportunity for fusion. We are also using a selective system in which hybrids will be isolated in soft agar as well as in monolayer. Third, we are trying to do the hybridization without Sendai virus and may eventually try lysolecithin.

ACKNOWLEDGMENTS

My collaborators in the work on erythrocytoid cells are Dr. Arthur Bank, Dr. Leslie Johnson, Mrs. Selina Kiang, Dr. Paul Marks, and Dr. Richard Rifkind.

I should like to acknowledge my co-author, Dr. Horst Brunschede, in the experiments on xanthine oxidase.

REFERENCES

1. Brunschede, H., and Krooth, R. S. (1973): Studies on the xanthine oxidase activity of mammalian cells. *Biochem. Genetics 8*:341–350.
2. Friend, C., Scher, W., Holland, J. G., and Sato, T. (1971): Hemoglobin synthesis in murine virus-induced leukemic cells *in vitro:* Stimulation of erythroid differentiation by dimethyl sulfoxide. *Proc. Nat. Acad. Sci. 68*:378–382.
3. Ostertag, W., Crozier, T., Kluge, N., Melderis, H., and Dube, S. (1973): Action of 5-bromodeoxyuridine on the induction of hemoglobin synthesis in mouse leukaemia cells resistant to 5-BUdR. *Nature New Biol. 243*:203–205.

DISCUSSION

Littlefield: Can you make heterokaryons with the erythroid cells? Even with heterokaryons, you might get some useful information about the regulation of hemoglobin production.

Krooth: It would be interesting to look at that, but we have not carried out our experiments in a way that would allow us to study heterokaryons.

Somatic Cell Hybridization, edited by
Richard L. Davidson and Felix de la Cruz.
Raven Press, New York © 1974.

Genetic Studies with Human Lymphoblast Cells

John Littlefield

Genetics Unit, Massachusetts General Hospital, Boston, Massachusetts 02114

I would like to discuss three areas of work that have recently been completed in my laboratory.

Daphne Kamely and Richard Erbe have been working on the regulation of the methionine biosynthesis pathway in BHK cells (1). In man, the major pathway of methionine biosynthesis is from homocysteine. The enzyme involved, a methyltransferase, is complicated and interacts with methyl-B_{12}, S-adenosyl methionine, and 5-methyltetrahydrofolate. We have confirmed that the B_{12} acts to either stabilize or activate the methyltransferase enzyme. We have also shown that the deprivation of methionine, the end product, results in a 2.5- to 4.0-fold increase in methyltransferase activity over a period of 1 to 2 days. This increase is blocked by puromycin, and probably represents a new example of end-product inhibition of synthesis. Relatively few such systems have been demonstrated firmly so far in animal cells.

The second item of work involves attempts by Ronald Zielke to synchronize human lymphoblast cultures (4). These are permanent lines of diploid, or near-diploid, immunoglobulin-producing cells that grow in suspension. We have worked with a wide variety of different blocking techniques but have focused primarily on excess thymidine. Despite much attention to the technical details for producing synchrony and attempts to minimize the toxicity of the block, in no line have we been able to get over 60% of the cells to divide in any cycle. We were unable to enrich these cultures for cells that can divide either by cloning or by repetitive synchronization up to 12 times in succession. There is no decrease in viability as judged by vital staining or cloning efficiency, and Epstein-Barr virus is not induced. We are left with the possibility that the excess thymidine block may somehow be inducing a temporary, nongrowing, G_0-like state.

Finally Stuart Orkin with Philip Buchanan and William Yount of Chapel Hill have been studying immunoglobulin production in mouse fibroblast–human lymphoblast hybrids (3). We fused thymidine kinase–deficient mouse cells (3T3-C2) with HGPRT-deficient human lymphoblasts (T5-1), produced by nitrosoguanidine mutagenesis. This line does not revert, has a normal diploid female karyotype, and produces lambda light chains. Hybrids were isolated after Sendai virus treatment and HAT selection, and thus far two hybrid clones have been studied extensively. They were obtained from independent fusion experiments. When first examined, each had a 2S mouse chromosome complement plus in one hybrid an average of 13 biarmed chromosomes and in the other 17 biarmed chromosomes, which were presumably human in origin. Both the hybrids continued to produce lambda light chain.

We have searched for other human immunoglobulin chains, and Colten has looked for human complement factors. We have also looked for mouse immunoglobulins or complement components. Thus far we have found only the human lambda light chain in the parental T5-1 lymphoblasts as well as in the two hybrids.

These mouse-human hybrid clones that continue to produce lambda chains resemble the mouse-mouse hybrids reported to continue to produce kappa chains (2). These experiments on continued gene expression in hybrid cells worry me, as do those on collagen and albumin synthesis. Conversely I am worried also about the constitutive or "household" enzymes which are occasionally extinguished in hybrids even though the genes can be shown to be present. I wonder if perhaps the separation into "luxury" and "household" functions needs to be expanded or if our classification of differentiated functions needs reexamination and qualification. Possibly proteins such as collagen, albumin, and immunoglobulins are more fixed in their control in contrast to inducible enzymes, so that they are less susceptible to extinction. They might form another category in between "luxury" and "household" functions.

REFERENCES

1. Kamely, D., Littlefield, J., and Erbe, R. (1973): Regulation of methyltetrahydrofolate: homocysteine methyltransferase activity by methionine, B_{12} and folate in cultured baby hamster kidney cells. *Proc. Nat. Acad. Sci. 70*:2585.
2. Mohit, B. (1971): Immunoglobulin G and free kappa-chain synthesis in different clones of a hybrid cell line. *Proc. Nat. Acad. Sci. 68*:3045.
3. Orkin, S., Buchanan, P., Yount, W., and Littlefield, J. (1973): Lambda chain production in human lymphoblast-mouse fibroblast hybrids. *Proc. Nat. Acad. Sci. 70*:2401.
4. Zielke, H., and Littlefield, J. (*in press*) Repetitive synchronization of human lymphoblast cultures with excess thymidine. In: Methods in Cell Physiology (ed. D. M. Prescott), Academic Press, New York.

DISCUSSION

Weiss: I wonder if one couldn't make a different separation. Proteins that are exported are in general not entirely extinguished, or at least there are almost always some cases in which they are not. This would include complement, albumin, gamma globulins, other serum proteins, and collagen.

Littlefield: That is a very good point.

Ephrussi: I do not think that the distinction between luxury and nonluxury functions is necessarily as clear-cut as I used to present. A variety of mechanisms, including dosage effects, may affect different systems in different ways. My conclusion is that we have to wait and see. As I said earlier, the situation will hopefully be clarified, but first it will seem to reach maximum complexity.

Pontecorvo: There are some recent data that suggest that exit from G_0 is a sort of stochastic process that a cell can go through within 2 min after entering G_0 or within 2 months *(personal communication).*

Littlefield: I do not know of that work. Certainly the ability to stop and wait in G_0 for a while has not been studied adequately and may be very important.

Siminovitch: What happens in the second cycle as far as that 60% of dividing cells?

Littlefield: We have done 12 repetitive cycles of synchronization and there are never more than 60% of the cells dividing in any cycle.

Siminovitch: What happens to the second cycle without a second block?

Littlefield: Synchrony is less and is lost within a cycle or two.

Somatic Cell Hybridization, edited by
Richard L. Davidson and Felix de la Cruz.
Raven Press, New York © 1974.

Differentiation and Somatic Cell Hybrids—A Commentary

Boris Ephrussi

Centre de Gentique Moleculaire, 91. Gif-sur-Yvette, France

All through my long scientific life I have been guided by the desire to build a bridge between genetics and embryology. It was with this idea that I began to work on cell hybridization when it was discovered. It is, therefore, a matter of rather great satisfaction to me, at least quantitatively, to see how the work initiated by Davidson and myself around 1964 has reached the present amplitude. In other words, I am very satisfied to see that so many people have begun working on the embryological problem defined as differentiation with the genetic tools.

As we have seen, those who have participated in this work have discovered a number of new and rather striking phenomena. The significance of the results (and here comes the question of qualitative appreciation) is much more obscure at the present time than it appeared when our first experiments were published. The situation was then very simple. Extinction was a simple rule. None of the present complications about which we have heard existed. Nevertheless, my long scientific past leads me to the optimistic conclusion that this is the usual course of events. You first discover some principle, then you find a number of exceptions, then the picture gets completely confused until one day there are some unexpected results that clarify and unify the field.

With this in mind, I would like to make a supplementary remark, and I would like to evaluate where the problems we have been handling really stand, where their position is in a much wider context of embryology. First, let me tell you of a conversation that took place between myself and André Lwoff in 1959 when I was planning to work via tissue culture into embryology. Cell hybridization hadn't been discovered yet, and I was looking for other genetic tools to approach the study of development. I wanted to go into *in vitro* systems, and I decided to go to California for 6 months to work with Dulbecco to learn the modern techniques of tissue culture and virology. Lwoff was just back from Dulbecco's lab where he acquired the techniques of tissue culture and was working on polio virus. Before going to Cal Tech, I visited Lwoff, who was an old friend, and I said to him, "What I want to do is start studying embryology using viruses as a tool for recognition of cell types." Lwoff listened to me, and then said, "This, I think, is a very sound idea, but what I think you should do is try to work with clones." I said to him, "André, don't you realize that an embryo *is* a clone."

What I am driving at is that really all the phenomena that we have been discussing, whatever their final significance, relate only to the final stages of differentiation. We cross differentiated cells that show one trait with cells that are often thought of as undifferentiated fibroblasts, which in itself is a mistake. In any case, what we do is to follow the fate of a certain number of functions in hybrids between contrasting cells. What is forgotten is (1) a technical matter which may be highly important, namely that most of the time we work with cells of permanent lines; (2) further, that we are really very far from the initial event, called determination by embryologists, when several cell lineages that have limited epigenetic capacities arise from a single egg (cell lineages whose potential of differentiation is already restricted to only a fraction of the total genetic information contained).

From this angle, one must admit that we still do not know how to go directly to this fundamental act of determination, which occurs much earlier than the terminal phenomenon of differentiation about which we have been talking. The fundamental question which was alluded to by Attardi when he talked about cell membranes really comes in here. To me, the major problem involves the original primary factors that cause the asymmetry in the egg, which ultimately results in the limitation of the differentiation capacities of the different cell lineages. This problem has not yet been attacked, but I think we should keep in mind that this is really the fundamental problem. Because we are handling permanent lines, whether or not they reflect normal differentiation, we are far from the initial crucial event, which may or may not have its primary causes in the nucleus.

To me it appears highly probable that all the necessary information for development is in the nucleus, simply because the whole program of development is transmitted from generation to generation. That does not mean, however, that the initial divergence is the result of the nucleus itself. It may be the result of the cell membrane, beginning with the membrane of the egg, not to speak of other cytoplasmic components. As you know, the old ideas of organ-forming and morphogenetic substances are coming into fashion again. For example, many people today are very much interested in the organ-forming systems in Drosophila.

Turning to the results of this session, the most important finding presented to us was that of reexpression. The reason is that reexpression shows the stability of what I call the epigenotype. It is not known whether reexpression is due uniquely to the intervention of the genes responsible for the particular differentiated function, although there is obviously correlation with chromosome loss, or whether there is also involvement of original surface components, which may propagate by cell heredity not necessarily dependent on the nucleus. This is a major question.

I will not go so far as to suggest, as Sonneborn did some time ago, that the cortex of the cell has its own complete phylogenetic development and is endowed with reproductive properties, but it is not excluded, in my opinion, that the presence of certain surface components determined by the nucleus is a prerequisite for the development of a differentiated line.

I repeat that we are still very far from being able to answer the question. The situation is very confused, but I look optimistically to the future.

Somatic Cell Hybridization, edited by
Richard L. Davidson and Felix de la Cruz.
Raven Press, New York © 1974.

Genetic and Biochemical Studies on Mouse Fibroblasts Bearing Mutations in Enzymes of the Purine and Pyrimidine Salvage Pathways

Teh-Sheng Chan*

Department of Biology, Massachusetts Institute of Technology, Cambridge, Massachusetts 02139

I would like to present some data on a number of different lines of work being carried out in the laboratory of Howard Green. These include studies on mutants in the salvage pathway enzymes for purines and pyrimidines and the phenotypes they present, together with their use in preparing and studying hybrid cells.

In one series of experiments, the linkage relations of the human genes for several enzymes of the purine and pyrimidine salvage pathways were studied with human-mouse hybrid cell lines (6).

Human-mouse hybrids were selected either in a hypoxanthine, aminopterin, and thymidine (HAT) medium or in an alanosine and adenine (AA) medium (7). The number of human chromosomes in all of these hybrids was reduced extensively; only one or a few metacentric chromosomes of human origin remained in the hybrids. Most of the human and mouse isozymes were separated by either isoelectric focusing or electrophoresis on a sheet of Cellogel. Human and mouse 5′-nucleotidase are membrane-bound and were separated by solubilizing the human enzyme by sonication and then by centrifugation at 40,000 × g to pellet the membrane-bound mouse enzyme. Table 1 summarizes the tests for the presence of several human enzymes in hybrid cell lines. Hybrids FAM-2, HM-C, and WE-28 were selected for human thymidine kinase (TK). They were shown to contain the human but not the mouse TK. DF8-V3A, selected in AA medium, contained only the human adenine phosphoribosyl transferase (APRT). Likewise, the hypoxanthine-guanine phosphoribosyl transferase (HGPRT) in T8-W2B and T8-W2C was confirmed to be of human origin.

As shown in the table, the three classes of hybrids contained those human enzymes for which selection was applied but none of those for which selection was not applied. We also tested for the presence of human adenosine

* Present address: Department of Physiology, University of Connecticut Health Center, Farmington, Connecticut 06032.

TABLE 1. *Human salvage pathway enzymes in human-mouse hybrid lines.*

		Tests for human enzymes of the purine salvage pathways					
Human-mouse hybrid	Parental cells	TK	APRT	HGPRT	AK	dCD	5′-Nucleotidase
FAM-2	Cl.1D (TK⁻)–FA	+			0		
HM-C	Cl.1D (TK⁻)–WI-38	+	0				0
WE-28	3T3-4E (TK⁻)–WI-38	+		0		0	
DF8-V3A	3T6-DF8 (APRT⁻)–FA		+	0	0	0	0
T8-W2B	3T6-TG8 (HGPRT⁻)–WI-38		0	+	0	0	+
T8-W2C	3T6-TG8 (HGPRT⁻)–WI-38		0	+	0		+
T8-W2B/TG				0			+
T8-W2C/TG				0			+

Modified from ref. 6. Enzyme deficiencies of the mouse parental cells are indicated in parenthesis. FA and WI-38 are human diploid fibroblast strains. +, Presence; 0, absence.

kinase (AK) and deoxycytidine deaminase (dCD) in those lines and the results were negative. Human 5′-nucleotidase was present in both T8-W2B and T8-W2C, suggesting that it was linked to HGPRT. However, thioguanine-resistant sublines of T8-W2B and T8-W2C both contained human 5′-nucleotidase although HGPRT was lost. We conclude, therefore, that the human genes for AK, dCD, and 5′-nucleotidase are not linked to any of the TK, APRT, and HGPRT genes. The consequence of these results is that AK, dCD, 5′-nucleotidase, and perhaps other salvage enzymes could be used as new selection markers for the construction of reduced human-mouse hybrids which retain other specific human chromosomes.

In another study, a mutant deficient in deoxycytidine deaminase was isolated from the mouse line 3T6 by using bromodeoxycytidine as a selection agent (3). The mutant contained about 0.5% the enzyme activity of 3T6 and had a reversion frequency of less than 10^{-7}. A fusion experiment with this mutant is in progress.

Purine excretion by an adenosine kinase–minus mutant has also been examined (2). The mutant was obtained from the 3T6 line by double selection using 6-methylthioinosine and then tubercidin. The mutant is a line with some unusual properties; when the mutant cells grew to a high density, they died within 15 hr. The medium harvested from a dying culture is toxic to wild-type cells, even if it is mixed 1:1 with fresh medium, suggesting that a toxic compound(s) is excreted into the medium by AK⁻ cells. We then studied the purine excretion of these AK⁻ cells by labeling the culture with (^{14}C)formate and identifying the excreted purines by thin-layer chromatography.

The results of these experiments are summarized in Table 2. Compared to the wild-type cells, the AK⁻ mutant, 3T6-TM, excreted 10 times more xanthine and 40 times more hypoxanthine. The HGPRT-deficient cells,

TABLE 2. *Labeled purine excretion by different lines*

	c.p.m. per mg cell protein			
	Wild type			
	3T6	3T3	Human diploid (SB)	
Xanthine	800	1,800	2,700	
Hypoxanthine	130	160	740	
Total	930	1,960	3,440	
	$HGPRT^-$			
	3T6-TG8	Human diploid K1	Human diploid 284	
Xanthine	47,100	30,300	38,000	
Hypoxanthine	29,200	53,300	53,200	
Total	76,300	83,600	91,200	
	AK^-			
	3T6-TM	3T6-TM2	3T6-TM3	3T6-TM4
Xanthine	9,100	4,300	3,800	15,900
Hypoxanthine	5,300	3,800	1,100	8,600
Total	14,400	8,100	4,900	24,500

From ref. 2.

either of mouse origin (3T6-TG8) or from human patients with the Lesch-Nyhan syndrome (KL and 284), excreted even more purines. These HGPRT-deficient cells grew normally. Therefore it seems unlikely that excessive amounts of hypoxanthine and xanthine alone are the reason for the toxicity of the culture medium of 3T6-TM. Probably the cells excreted some other compounds that are not detectable by our method and are toxic to cultured cells.

The excretion of purines by $HGPRT^-$ cells is due to a defect in the enzyme, which reutilizes hypoxanthine and guanine by converting them into nucleotides. An explanation for the excretion by the AK^- cells is not so straightforward because the cells have the HGPRT to reutilize hypoxanthine and guanine. Balis (1) noticed in the purine salvage pathways the possibility of a cyclic conversion of adenosine to AMP through inosine, hypoxanthine, and IMP (Fig. 1). The enzyme adenosine kinase catalyzes phosphorylation of adenosine to convert it directly to AMP. Thus Green and Ishii (4) postulated that the role of the kinase was to prevent the cycle from operating excessively. In the absence of AK, adenosine formed by 5′-nucleotidase action on AMP would be fed into the cellular hypoxanthine pool. Some of the hypoxanthine would be reutilized by the action of HGPRT, and some would be lost into the medium. The data presented above support this hypothesis. The observation has clinical significance in that some forms of human gout could be caused by a hereditary deficiency of adenosine kinase.

I would also like to describe briefly some work done in our laboratory on

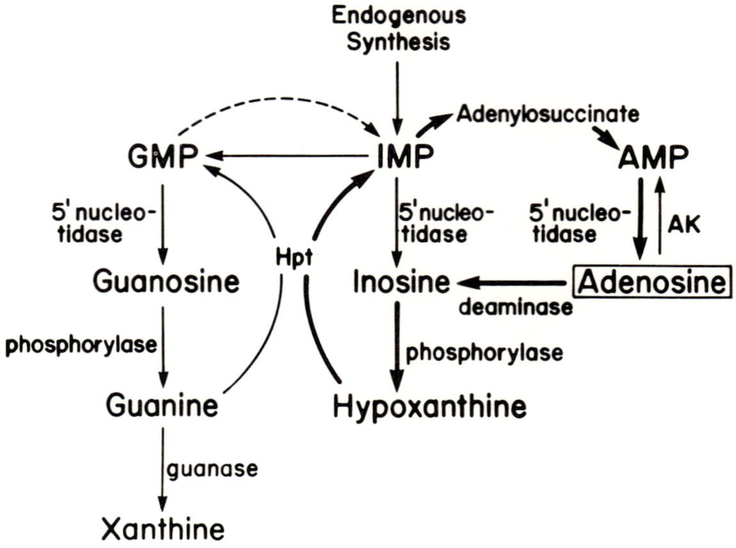

FIG. 1. Purine salvage pathways. Some intermediates are omitted. Heavy arrows indicate adenosine cycle. (From ref. 2.)

the toxicity of adenosine to cultured mammalian cells (5). Calf serum and fetal calf serum contain adenosine deaminase activity, whereas horse serum does not. When horse serum is used in the culture medium, adenosine is found to be toxic to the mouse fibroblast 3T6 at concentrations as low as 5×10^{-6} M. At 2×10^{-5} M all the cells are killed. The toxicity of adenosine seems to require the activity of AK because the AK^- derivative of 3T6 is resistant to a 70-times higher concentration of adenosine. The toxicity of adenosine can be prevented by the addition of uridine, cytidine, deoxyuridine, deoxycytidine, or uracil to the growth medium. Adding orotate, orotidine, cytosine, thymidine, or any of the purines has no effect. When cells are treated with adenosine, the cellular pool of uridine nucleotides decreases as much as 70%. On the other hand, orotate in the cells and in the medium increases about sevenfold. It appears that adenosine or more likely, one of its nucleotides, exerts its toxic effect on cultured mammalian cells by interference with pyrimidine biosynthesis.

REFERENCES

1. Balis, M. E. (1968): Aspects of purine metabolism. *Fed. Proc. 27*:1067.
2. Chan, T.-S., Ishii, K., Long, C., and Green, H. (1973a): Purine excretion by mammalian cells deficient in adenosine kinase. *J. Cell. Physiol. 81*:315.
3. Chan, T.-S., Long, C., and Green, H. (1973b): A deoxycytidine-cytidine deaminase deficient mutant of a mouse fibroblast. *Fed. Proc.* Abstr. 566.
4. Green, H., and Ishii, K. (1972): On the existence of a guanine nucleotide trap, the role of

adenosine kinase and a possible cause of excessive purine production in mammalian cells. *J. Cell Sci. 11*:173.
5. Ishii, K., and Green, H. (1973): Lethality of adenosine for cultured mammalian cells by interference with pyrimidine biosynthesis. *J. Cell Sci. 13*:429.
6. Long, C., Chan, T.-S., Levytska, V., Kusano, T., and Green, H. (1973): Absence of demonstrable linkage of human genes for enzymes of the purine and pyrimidine salvage pathways in human-mouse somatic cell hybrids. *Biochem. Genet. 9*:283.
7. Kusano, T., Long, C., and Green, H. (1971): A new reduced human-mouse somatic cell hybrid containing the human gene for adenine phosphoribosyl transferase. *Proc. Nat. Acad. Sci. USA 68*:82.

DISCUSSION

Krooth: It is of some interest that we found in 1964 that diploid cell strains from patients with orotic aciduria were unable to grow in the presence of adenosine unless they were given uridine or cytidine [Krooth, R. (1964): *Cold Spr. Harb. Symp. Quant. Biol. 29*:189]. We then looked for the same effect in normal cells. We did the experiment in both human sera and calf sera and we found no effect. Ishii and Green have now demonstrated that there is an adenosine deaminase present in sera of both species.

Somatic Cell Hybridization, edited by
Richard L. Davidson and Felix de la Cruz.
Raven Press, New York © 1974.

Mechanisms of *De Novo* Variation* in Mammalian Cell Cultures

Morgan Harris

Department of Zoology, University of California, Berkeley, California

The work I want to discuss deals with the nature of so-called "mutations" in mammalian cell cultures as seen, for example, in acquired resistance to specific drugs. Such variants can be readily obtained from established or permanent cell lines in which they arise by spontaneous change and often in relatively high frequency (see ref. 2). We found that *de novo* variants can be obtained from polyploid cells as well as from diploid cells and several years ago used this observation to compare mutation rates in Chinese hamster cells at different ploidy levels (3, 4). As determined by fluctuation tests, the mutation rates for two different markers, temperature sensitivity and resistance to 8-azaguanine, seemed to remain constant or decline only slightly in cells that were diploid, tetraploid, and octaploid respectively.

These observations have been difficult to understand in terms of classic genetic models. Some uncertainty existed, however, since we did not know whether heat sensitivity and azaguanine resistance behaved as dominants or recessives in this model system. For further study we have now isolated two new markers and have characterized their mode of phenotypic expression (5). These variations involve resistance to vinblastine and to cytosine arabinoside (ara-C) respectively. We can isolate variants that are resistant to ara-C or vinblastine from mass populations of cells which are, in addition, deficient in HGPRT or thymidine kinase activity. The latter properties may then be used to construct hybrids in HAT medium without selective pressure on the markers under investigation.

When such hybrids were isolated from a cross between vinblastine-sensitive and -resistant cells, the hybrids showed the same degree of resistance to vinblastine as did the resistant parent cell (Fig. 1). Hence vinblastine resistance is dominant in phenotypic expression. In contrast, when ara-C sensitive cells were crossed with ara-C resistant cells, the hybrids were as sensitive to the drug as were the sensitive parent cells (Fig. 2). Thus ara-C resistance behaves as a recessive marker at the phenotypic level.

On the basis of these findings we designed experiments to compare the

---

* Supported by grant CA 12130 from the National Cancer Institute, U.S. Public Health Service and by Cancer Research Funds from the University of California.

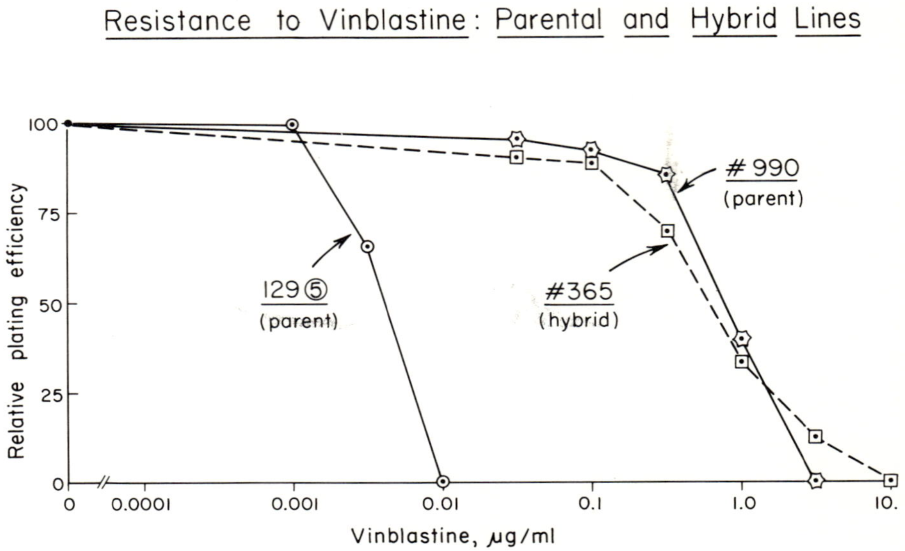

FIG. 1. Relative plating efficiency in vinblastine of sensitive line 129, resistant line 990, and hybrid line 365 derived from a cross between 129 and 990 cells. (From ref. 5.)

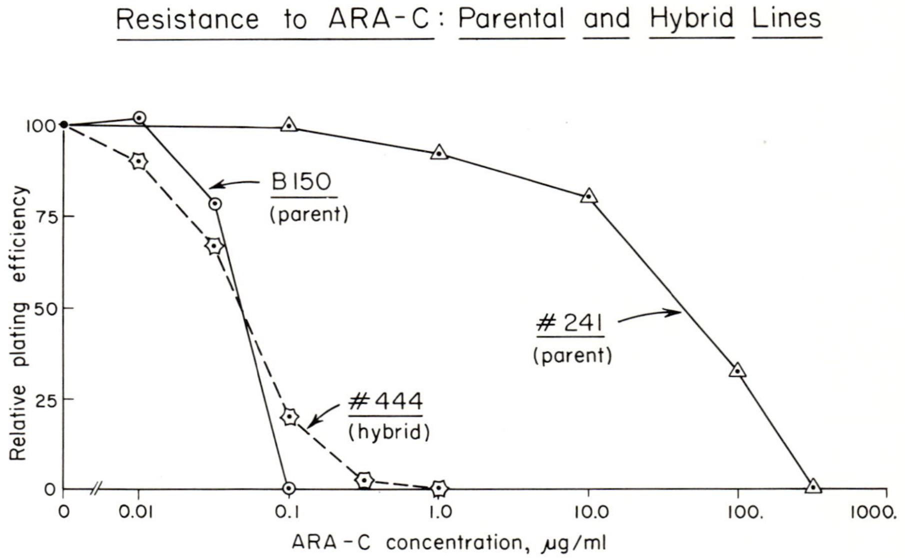

FIG. 2. Relative plating efficiency in ara-C of sensitive line B150, resistant line 241, and hybrid line 444 derived from a cross between B150 and 241 cells. (From ref. 5.)

frequency of induced variation for dominant and recessive characters respectively (7). Chinese hamster cells that were sensitive to vinblastine and to ara-C were used as a test system, and drug resistance was induced by a standard 2-hr treatment with nitrosoguanidine (MNNG). As in earlier experiments with azaguanine and heat resistance (6, 8), the cells were cultivated in normal medium for 6 days after mutagen treatment to allow for expression of variant characteristics. The derivative populations were then assayed for the frequency of resistant cells by plating in the presence of ara-C or vinblastine. With both drugs, the proportion of mutants rises markedly after exposure to MNNG. The rate of increase in variant colonies is approximately exponential with mutagen concentration when the number of survivors is corrected for differences in plating efficiency in the absence of drug.

It is instructive to study the frequency of induced mutations for both drugs on a single cell type. Such experiments are particularly clear-cut with tetraploid cells since the differential in expression of dominant and recessive markers should be accentuated. Figure 3 illustrates a comparison of this kind. Depending on the concentration of drugs used for assay, the curve for vinblastine resistance is observed to lie above that for ara-C resistance, or vice versa. But the important point is that there is no significant difference in the incidence of resistant colonies obtained by selection with the two drugs, either in untreated populations or in cultures exposed to graded levels of MNNG. Actually, the kinetics of response to mutagen treatment are essentially the same for the two markers (within the limits of experimental error the curves are roughly parallel).

These are not the expected results. For a dominantly expressed marker such as vinblastine resistance, each mutation should be expressed directly in variant cells. However, if the marker is a recessive one (e.g., ara-C resistance) phenotypic expression in tetraploid cells would require simultaneous mutation of the two or more genes concerned and/or loss of the unmutated wild-type allele(s) by somatic segregation. The net frequency of expression of the recessive marker, based on the product of the probabilities for these dual events, should be significantly less than the mutation rate *per se*. However, data show that the rates of expression of vinblastine- and ara-C resistance are in fact indistinguishable in normal or mutagenized populations.

Segregation studies on hybrids offer an alternative insight into the basis for *de novo* variation in cultures of Chinese hamster cells. The hybrids used originated by spontaneous fusion between cells deficient in HGPRT and thymidine kinase activity and were isolated as clones in HAT medium. We then attempted to segregate cells with the parental phenotypes by back selection of the hybrids in either azaguanine or BUDR. As shown in Table 1, for independently derived hybrids, the incidence of segregants varies considerably. For resistance to azaguanine the figures range from 0.2 to 2%, and successive assays with the same hybrid clone gave approximately the

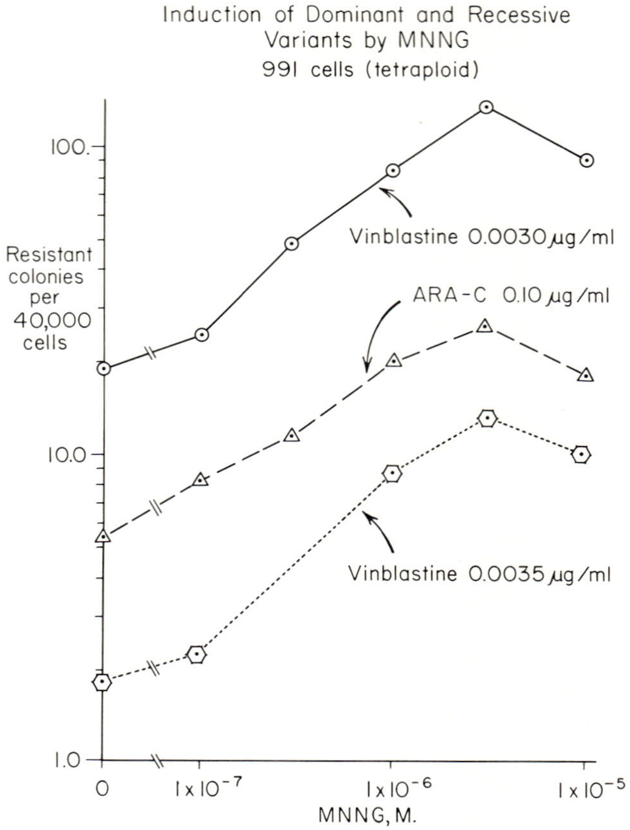

FIG. 3. Comparison of dominant (vinblastine) and recessive (ara-C) mutations to drug resistance after MNNG treatment of tetraploid populations of Chinese hamster cells.

TABLE 1. *Segregation of variants from hybrid Chinese hamster cells*[a]

	Segregants per 100 cells in selective media	
Hybrid line	Azaguanine 30 μg/ml	Bromodeoxyuridine 30 μg/ml
272	2.1	1.6
579	0.2	4.3
599	0.25	2.0
600	0.8	7.5

[a] All hybrids were isolated as clones from mixtures of line 129 ($HGPRT^-$) X line B150 (TK^-) in HAT medium. (From ref. 7.)

same frequency of segregants. The incidence of BUDR-resistant segregants is roughly similar and also varies significantly between different hybrid lines. If segregation depends on chromosome loss or rearrangements, one would expect the frequency of BUDR-resistant segregants to be significantly lower than the incidence of azaguanine-resistant segregants. This follows from evidence that suggests that the locus for azaguanine resistance in Chinese hamster cells (10) is on the X chromosome as in man, whereas BUDR resistance, in human cells at least, is an autosomal marker (9). Thus, hybrids would be expected to lose two chromosomes to expose the BUDR resistance marker, but only one chromosome to confer azaguanine resistance. However, the BUDR segregants were at least as numerous as the azaguanine-resistant segregants.

To measure the actual rates of segregation, we have carried out fluctuation tests on colonies derived in each case from a clone of hybrid cells. In separate experiments on azaguanine and BUDR resistance we selected 15 independent colonies and determined the incidence of variants for each colonial population at two different cell densities. We observed a high rate of segregation for azaguanine resistance, between 3×10^{-2} and 4×10^{-3}. However, the segregation rate for BUDR resistance was not noticeably less, with tests at two cell densities giving values of 3.7×10^{-3} and 2.4×10^{-3} respectively. Rates of this magnitude strain the credibility of chromosomal changes as an implementing mechanism. Since in these hybrids there are approximately 44 chromosomes, it would be necessary if loss is random to postulate that one chromosome per cell is altered or lost at each division in order to account for segregation frequencies of 2% or more. We have not yet made extensive chromosome studies on our segregant lines, but limited counts give no evidence of extensive loss. This is consistent with the observations of Chasin (1) who isolated segregant lines of Chinese hamster cells by back selection of hybrids in thioguanine. Chasin found the average of the modal chromosome numbers for 10 segregant sublines to be 39, identical with that of the original hybrid line.

I would also like to discuss one very puzzling observation in this segregation system. Of the clones isolated by back selection in azaguanine or BUDR, 80 to 90% or more retain in varying degree the ability to grow also in HAT medium. Only 10 to 20% of the segregants are clearly HAT sensitive. It is difficult to account for the ability of these intermediate cell types to grow in HAT as well as azaguanine or BUDR. The azaguanine-resistant parental cell lacks HGPRT activity and the frequency of reversion is less than 10^{-7}. Similarly, the BUDR-resistant parental cell lacks thymidine kinase activity and has a reversion frequency of 10^{-6} or less. If the recessive $HGPRT^-$ or TK^- alleles were merely being exposed by chromosome segregation, all of the segregants would show the parental (HAT sensitive) phenotype, which they do not do. Hence, chromosome segregation does not explain these results. Another possible explanation might be the occurrence of mutation at the $HGPRT^+$ or TK^+ loci received from the opposite parental cell. To investigate this possibility, we measured the frequency at which azaguanine-resistant colonies arise in the BUDR-resistant parent line. Such cells occur with a frequency of about 10^{-4}, but this is far lower than the segregation frequencies actually observed in hybrids. All attempts to secure BUDR-resistant variants by one-step selection from the azaguanine-resistant line failed. Therefore, mutation at the $HGPRT^+$ or TK^+ alleles does not seem to explain our back-selection data.

In conclusion, our observations on acquired drug resistance are difficult to reconcile with the usual explanations based on gene mutation and chromosomal change. It seems possible that in these cell lines, at least, the underlying mechanisms involve alterations in gene expression rather than changes in genetic information. The frequency of such shifts, if they occur, appears to be greatly enhanced in hybrids as compared to parent cells. This suggests that specific alterations in phenotypic expression may in some way be made more probable in the hybrid cells because one of the parental cells has a corresponding enzyme deficiency. Further studies now in progress should enable us to examine and evaluate this hypothesis.

REFERENCES

1. Chasin, L. A. (1972): Non-linkage of induced mutations in Chinese hamster cells. *Nature New Biol. 240*:50–52.
2. Harris, M. (1964): *Cell Culture and Somatic Variation,* Holt, Rinehart and Winston, New York.
3. Harris, M. (1971a): Polyploid series of mammalian cells. *Exp. Cell Res. 66*:329–336.
4. Harris, M. (1971b): Mutation rates in cells at different ploidy levels. *J. Cell Physiol. 78*:177–184.
5. Harris, M. (1973a): Phenotypic expression of drug resistance in hybrid cells. *J. Nat. Cancer Inst. 50*:423–429.
6. Harris, M. (1973b): Anomalous patterns of mutation in cultured mammalian cells. *Genetics 73*:181–185.
7. Harris, M. *(in press):* Model systems for evaluation of mutations in cell cultures. *Int. Rev. Cytol., Suppl.*
8. Harris, M. *(in press):* Induction of variants by nitrosoguanidine in diploid and tetraploid lines of Chinese hamster cells. In: *World Symposium on Model Studies in Chemical Carcinogenesis,* Marcel Dekker, New York.
9. Miller, O. J., Allderdice, P. W., and Miller, D. A. (1971): Human thymidine kinase gene locus: assignment to chromosome 17 in a hybrid of man and mouse cells. *Science 173*:224–245.
10. Westerveld, A., Visser, R. P. L. S., and Freeke, M. A. (1971): Evidence for linkage between glucose 6-phosphate dehydrogenase and hypoxanthine-guanine-phosphoribosyl transferase loci in Chinese hamster cells. *Biochem. Genetics 5*:591–599.

DISCUSSION

Littlefield: I think that this sort of work is very useful in reminding us of the confusion that exists in terms of the meaning of variation in heteroploid lines.

Siminovitch: I know there is not very much value in negative results, but

Mike McBurney and George Whitmore in our laboratory have tried to isolate ts mutants from a tetraploid line and have never succeeded under conditions in which we always succeed with the diploid cells.

Harris: We have had no difficulty isolating stable variants of several kinds from tetraploid Chinese hamster cells, including mutants resistant to azaguanine and to ara-C, both of which behave as recessives. The temperature-sensitive phenotype is also recessive in your mutants, is it not?

Siminovitch: Yes.

Miller: I am always very careful about dismissing chromosome change as a possible explanation without actually looking at the chromosomes. When we have used banding techniques to study various lines of established cells, mainly mouse or rat, we find an enormous amount of variability. Therefore, a level of one or two chromosomal changes per mitosis would be quite in line with my expectation. I am not as convinced as you are that chromosome segregation can be dismissed as an explanation for these high rates of resistance.

Harris: How do you explain Chasin's results? In 10 drug resistant sublines, he found exactly the same modal number as in the original hybrid.

Miller: Banding studies on Chinese hamster lines indicate that there is a tremendous amount of variation within the chromosome complement. The fact that the number is constant is therefore meaningless. This is true in mouse lines as well. I feel strongly that one can not say what is happening to chromosomes without actually looking at them.

Harris: I agree that observations on chromosome number alone will not resolve the problem. The main issue, however, is the unacceptably high rate of chromosome change in hybrid cells that must be assumed in order to account for the frequencies observed for segregation from hybrids. Depending on one's viewpoint, a loss of one or two chromosomes from each hybrid cell at mitosis, balanced in some way (e.g., by nondisjunction) to preserve constancy of chromosome number, is perhaps of borderline credibility. But what about the BUDR-resistant segregants? These are at least as numerous in our material as azaguanine-resistant segregants. Since the genetic loci for BUDR resistance are presumably autosomal rather than on the X chromosome (for azaguanine resistance) the frequency of chromosome alteration or loss would have to be squared. Such an assumption would seem to be an unrealistic one.

I would like to make one more comment. No one is questioning that gene mutation occurs in mammalian cells as in all other genetic systems. The issue is whether *de novo* variations such as drug resistance in cell cultures are necessarily the result of gene mutation. At least some of these variants show properties that are not consistent with present models of gene mutation. For this reason we should retain an open mind to other possible mechanisms. Many such variations when better analyzed may turn out to represent changes in gene expression rather than alterations in genetic information.

Somatic Cell Hybridization, edited by
Richard L. Davidson and Felix de la Cruz.
Raven Press, New York © 1974.

Isolation and Characterization of Mutants of Somatic Cells

Louis Siminovitch

Departments of Medical Genetics and Medical Biophysics, University of Toronto, The Ontario Cancer Institute, and The Hospital for Sick Children, Toronto, Canada M5S 1A8

I will try to provide an overview of the work in our laboratory on the generation and characterization of different types of mutants in somatic cells. We are interested in developing a battery of mutants for use in genetic studies of structure and function in somatic cells.

We began our work with the mouse L-cell line but most of our recent work has been done with Chinese hamster ovary (CHO) cells. As described by Puck, CHO cells grow well in suspension, clone well, and have a fairly stable karyotype. One major reason for using these cells is that they permit the isolation of auxotrophic mutants, and their consequent use as markers and in hybridization experiments. We have looked at two different classes of mutants: temperature-sensitive (ts) mutants, which grow at 34°C but not at 38°C, and drug-resistant mutants.

For the isolation of ts mutants, we have used a selection system that is different from that described by Puck but similar in principle. Instead of using BUdR and light, we have used high-specific activity tritiated thymidine to kill dividing cells (8, 9). We are now able to use one cycle or at most two or three cycles of suicide for the selection. We have obtained a large number of mutants, most of which have not been characterized in depth. One ts mutant, TsHl, has been so characterized, however (6, 7). The mutant is a clean one, in the sense that its plating efficiency is down to 10^{-7} at the nonpermissive temperature. It can revert, and the reversion frequency to the wild type is increased by mutagens. It is of interest that this mutant is killed very rapidly when shifted from 34 to 38.5°C. When protein synthesis is measured at the nonpermissive temperature, the incorporation of radioactive amino acids is almost totally arrested after 15 min. Reversal of phenotype is also rapid; if the cells are shifted from 38.5 to 34°C, protein synthesis returns to normal in about 20 min.

Stanners and Thompson have looked at the ribosome profile in those cells, and they have observed a very rapid shift from polysomes to monosomes at the nonpermissive temperature (6). They have recently shown that this mutant is characterized by a temperature-sensitive leucyl t-RNA synthetase (7).

We have also been looking at three or four different types of drug-resistant mutants. One of these is a mutant resistant to α-amanitin, a drug which inactivates RNA polymerase-II (10). Mutants that are resistant to 1 μg/ml of α-amanitin can be selected from CHO cells (4). There appear to be two classes of such resistant mutants; one class in which RNA polymerase-II is altered so that it is no longer inhibited by the drug, and one that may involve changes in permeability.

The enzyme has been partially purified by DEAE cellulose chromatography in the first type of mutant. The purified enzyme is resistant to α-amanitin inactivation, but the level of the enzyme in the cells does not seem to be changed significantly. Since the characteristics of the enzyme are altered, we have hypothesized that the mutation involves a structural gene (4).

A second type of drug-resistant mutant studied involves resistance to ouabain (1, 2). Ouabain is a cardioactive steroid that affects the membrane sodium-potassium activated ATPase. Mutants resistant to ouabain can be selected easily from hamster or L cells, and this has proven to be a useful system from the genetic point of view. In contrast to selection with drugs such as 8-azaguanine, all the wild-type cells are quickly eliminated by ouabain and colonies of resistant cells are seen against a clean background. The mutants are stable and do not revert in the absence of the selective agent (2).

Baker and Mankovitz (2) have used cell hybridization to test the dominance or recessiveness of the ouabain resistant mutation. The cell lines used for the isolation of the drug-resistant mutants already carried auxotrophic markers and the hybrids were selected using these markers rather than the HAT-selection system. Cells auxotrophic for one set of nutrients were crossed with cells auxotrophic for other nutrients and hybrids that require none of these nutrients were selected. One parental line was ouabain sensitive and the other had been selected for ouabain resistance. In such crosses, the number of hybrids obtained is the same whether or not ouabain is present in the selective medium. We have concluded that ouabain resistance is a dominant or codominant mutation.

A third mutation studied in our laboratory involves resistance to colchicine (5). In order to obtain mutants affecting the spindle proteins, Ling and Thompson began to isolate mutants resistant to colchicine. These were tested for binding of colchicine to the spindle protein. None of the mutants showed any alteration in the binding of colchicine to the spindle protein, but they did have marked changes in their ability to take up colchicine. The isolates seem, therefore, to be permeability mutants. Ling and Thompson (5) have obtained lines with a spectrum of levels of colchicine resistance, and the level of resistance can be correlated with the amount of uptake of the drug.

It is interesting that the colchicine-resistant mutants are pleiotrophic. When mutants resistant to colchicine are isolated, they also show an acquired resistance to a number of other substances such as actinomycin D (which has absolutely no structural relationship to colchicine), colcemid, and vinblastine. It would appear that the change in the membrane giving rise to colchicine resistance is responsible for the resistance to the other compounds. The same phenomenon is observed when actinomycin D resistant cells are isolated (3). This may thus be a general phenomenon; when the membrane is altered so that resistance is seen to one agent, its permeability to the other agents is simultaneously affected.

We have also isolated mutants resistant to concanavalin A (conA) (11). Such cells are easy to isolate and, in this case also, there are accompanying collateral changes. The cells become more sensitive to dibutryl-cyclic AMP and also to temperature. Since conA is probably acting at the membrane of the cells, we are thus observing another type of collateral change in the membrane.

In summary, our studies indicate that temperature-sensitive mutants can be obtained that are valuable for studying cell function. The studies on drug resistance have indicated that mutants that affect intracellular functions and the cell membrane can also be isolated.

REFERENCES

1. Baker, R. M., and Till, J. E. (1971): Isolation of ouabain-resistant mutants of mouse cells in tissue culture (Abstr.). *Fifteenth Annual Meeting of the Biophysical Society,* New Orleans, p. 283a.
2. Baker, R. M., Brunette, D. M., Mankovitz, R., Thompson, L. H., Whitmore, G. F., Siminovitch, L., and Till, J. E. (1974): Ouabain-resistant mutants of mouse and hamster cells in culture. *Cell 1*:9–21.
3. Biedler, J. L., and Riehm, H. (1970): Cellular resistance to actinomycin D in Chinese hamster cells *in vitro:* Cross-resistance, radioautographic, and cytogenetic studies. *Cancer Res. 30*:1174–1184.
4. Chan, V. L., Whitmore, G. F., and Siminovitch, L. (1972): Mammalian cells with altered forms of RNA polymerase II. *Proc. Nat. Acad. Sci. USA 69*:3119–3123.
5. Ling, V., and Thompson, L. H. *(in press):* Reduced permeability in CHO cells as a mechanism of resistance to colchicine. *J. Cell. Physiol.*
6. Stanners, C. P., and Thompson, L. H. (1974): The studies on a mammalian cell mutant with a temperature-sensitive leucyl-rRNA synthetase. In: *Control of Proliferation in Animal Cells,* Cold Spring Harbor, New York.
7. Thompson, L. H., Harkins, J. L., and Stanners, C. P. (1973): Identification of a conditional lethal mutation affecting leucyl-tRNA synthetase in CHO cells. *Proc. Nat. Acad. Sci. USA 70*:3094–3098.
8. Thompson, L. H., Mankovitz, R., Baker, R. M., Till, J. E., Siminovitch, L., and Whitmore, G. F. (1970): Isolation of temperature-sensitive mutants of L-cells. *Proc. Nat. Acad. Sci. USA 66*:377–384.
9. Thompson, L. H., Mankovitz, R., Baker, R. M., Wright, J. A., Till, J. E., Siminovitch, L., and Whitmore, G. F. (1971): Selective and nonselective isolation of temperature-sensitive mutants of mouse L-cells and their characterization. *J. Cell. Physiol. 78*:431–439.
10. Wieland, T. (1968): Poisonous principles of mushrooms of the genus amanita. *Science 159*:946–952.
11. Wright, J. A. (1973): Evidence for pleiotrophic changes in lines of Chinese hamster ovary cells resistant to concanavalin A and phytohemagglutinin-P. *J. Cell Biol. 56*:666–675.

DISCUSSION

Wiener: Have you hybridized the colchicine resistant lines?

Siminovitch: The results of Ling's experiments on the preliminary tests of hybrids with the colchicine resistant mutants are inconclusive. We have not yet made hybrids with the α-amanitin or conA resistant cells. The ts mutations all seem to be recessive.

Attardi: Is there an increase in the ouabain resistance of the ATPase in the ouabain resistant mutants?

Siminovitch: Yes.

Harris: We have found three drug resistance markers in Chinese hamster cells that behave as dominants in the manner you described. One of them is vinblastine resistance. Another is chloramphenicol resistance and a third is ethidium bromide resistance. There is some cross-resistance between vinblastine and ethidium bromide resistance, but none whatsoever with chloramphenicol. I think there are more of those dominantly expressed markers than had been realized in earlier work.

Siminovitch: I suspect that many dominant mutations will be found. I also want to make it clear that you may not see collateral effects every time you select a membrane mutant. Ouabain resistance does not seem to show any collateral effects, for example. I think that different membrane mutants will fall into different classes. It is of interest also that with colchicine resistance, one can obtain several different levels of resistance and these seem to be stable.

Pontecorvo: With mutations other than the α-amanitin resistance, is there evidence for a true genetic change?

Siminovitch: There is some evidence of mutants involving an HGPRT deficiency. In some azaguanine resistant cells that do not show HGPRT activity, there is some cross-reacting material. [Beaudet, A., Roufa, D., and Caskey, C. (1973): *Proc. Nat. Acad. Sci. 70*:320].

Gerald: What happens if you treat human cells with ouabain?

Siminovitch: HeLa cells are very sensitive to the drug. HeLa cells should be useful for study because the cells are much more sensitive than L cells or CHO cells and thus one can examine the binding of the drug in such cells and try to correlate this with resistance. Diploid human fibroblasts are very sensitive to ouabain, but I cannot give you an exact indication of the sensitivity.

Thompson: Have you examined ribosomal RNA synthesis in the presence of α-amanitin in the resistant cells?

Siminovitch: No. We have been concerned thus far only with alterations in the sensitivity of the polymerase to the drug. Lobban in our laboratory will be looking at the binding of radioactive α-amanitin to the polymerase.

Thompson: I asked the question because, although ribosomal RNA polymerase is resistant to the drug in cell-free systems, ribosomal RNA synthesis *in vivo* can be blocked by the α-amanitin.

Siminovitch: We have not done any tests in that area.

Chan: Has anyone looked for ts revertants from drug-resistant mutants?

Siminovitch: We have not done this systematically. This should be especially easy with the conA resistant cells because they are temperature sensitive. In that particular system it should be possible to go in both directions.

Somatic Cell Hybridization, edited by
Richard L. Davidson and Felix de la Cruz.
Raven Press, New York © 1974.

Temperature Sensitive Mutants of BHK-21 Cells

Claudio Basilico

Department of Pathology, New York University Medical Center, School of Medicine, New York, New York 10016

I plan to discuss the work we have been doing in the past two years on the isolation of temperature-sensitive mutants of BHK cells. This work has been done in collaboration with H. K. Meiss and D. Toniolo. The BHK Syrian hamster line is pseudodiploid and contains 44 chromosomes. In addition to the obvious uses of conditional lethal mutants for cell genetics and physiology, we are interested in tumor viruses, and BHK is a cell that eventually one could use for studying cell-virus interactions.

Our method of isolating mutants is essentially similar to that of Siminovitch. We mutagenize the cells, shift them for a period of time to the nonpermissive temperature (39°C), and try to kill the wild-type cells using fluorodeoxyuridine. FUDR should select for temperature-sensitive (ts) mutants affected in DNA synthesis but its specificity is not very good. That applies also to other selective agents. However, reconstruction experiments show that enrichment for the mutants does occur with the use of this selective agent.

After two or three temperature shift–FUDR cycles, we isolate surviving colonies and test the phenotype at 33 and 39°C. We generally find three types of mutants. Mutants of the first type have a high frequency of backmutation so that the efficiency of plating at 39°C relative to that at 33°C is on the order of 10^{-2} to 10^{-3}. These mutants have a high reversion rate, and they are difficult to work with. Thus, they were not analyzed further.

A second type of mutant seems to have some density dependent leakiness. This type will give no colonies if plated at 39°C at low density, but at high density will grow almost like wild type. This behavior could be due to a leaky mutation that is somewhat counteracted by cross-feeding effects. These mutants also were not analyzed further.

The third type of mutants we obtained were "good" mutants in that they had a reversion frequency of less than 10^{-6} and were not leaky. We now have 30 to 40 mutants that share these general properties. The phenotype after the temperature shift varies from mutant to mutant. Some will grow for about one generation after being shifted to 39°C, whereas others stop almost immediately.

We tested the various mutants for complementation by using cell hybridization. The important point was to establish whether or not the mutations were recessive. Since we were dealing with diploid lines, one might have thought that the only type of temperature-sensitive mutants one would get would be dominant lethals. This does not seem to be the case. For all the mutants we have crossed, the hybrids have been able to grow at 39°C. Standard complementation tests reveal that there is more than one complementation group. Most mutants belong to different complementation groups but a few groups contain more than one mutant (1).

One objection to the above experiment was that, if a dominant mutation was involved, we might have selected for hybrids that had lost the chromosome containing the mutant gene by isolating the hybrids at 39°C. To prove that this was not the case, we crossed ts HGPRT$^-$ with ts$^+$ TK$^-$ BHK cells. Thus, we could select the hybrids in HAT medium at 33°C, so there would be no selective pressure for the loss of a hypothetical temperature-sensitive dominant mutation (see Table 1). In this cross, the same frequency of hybridization occurred at 33° and 39°. Furthermore, hybrids selected at low temperature were able to grow when shifted to high temperature. Thus the mutations to temperature sensitivity seem to be recessive.

TABLE 1. *Hybridization between ts, HGPRT$^-$ and ts$^+$, TK$^-$ BHK cells*[a]

ts, HGPRT$^-$ parent	Number of colonies[b] at 39°C in HAT medium	Number of colonies[b] at 33°C in HAT medium	% colonies that continued growth upon shift from 33 to 39°C[c]	Hybrid clones isolated at 33°C and capable of growth at 39°C
T15	416	355	>90	4/4
T23	195	272	>90	3/3
T22	220	280	>90	5/5

[a] Each of the ts mutants was fused with thymidine kinase deficient BHK cells. After 2 days cultivation at 33°C, the cells were plated at 2×10^5/100mm Petri dish at the desired temperature in medium containing HAT (10^{-5} M aminopterin, 4×10^{-5} M thymidine, 10^{-4} M hypoxanthine, 10^{-5} M glycine).
[b] Out of 8×10^5 cells plated.
[c] Determined 4 to 5 days after shift.

In the last part of my talk, I will briefly describe the characterization of the ts 422E mutant, which is one of the so-called "good" mutants in that it has a low reversion frequency and is not leaky. When the cells are shifted to 39°C, cell division stops after about one generation (12 hr). However, the cells become progressively larger with time after the shift. Macromolecular synthesis is not significantly inhibited. Besides becoming increasingly larger, the nucleoli aggregate into one big mass in the nucleus. We thought this might be due to an accumulation of ribosomal RNA precursors in the nucleoli, so we examined the production of ribosomal RNA (see Fig. 1). Production of rRNA in the wild-type cells is essentially the same at 33 and

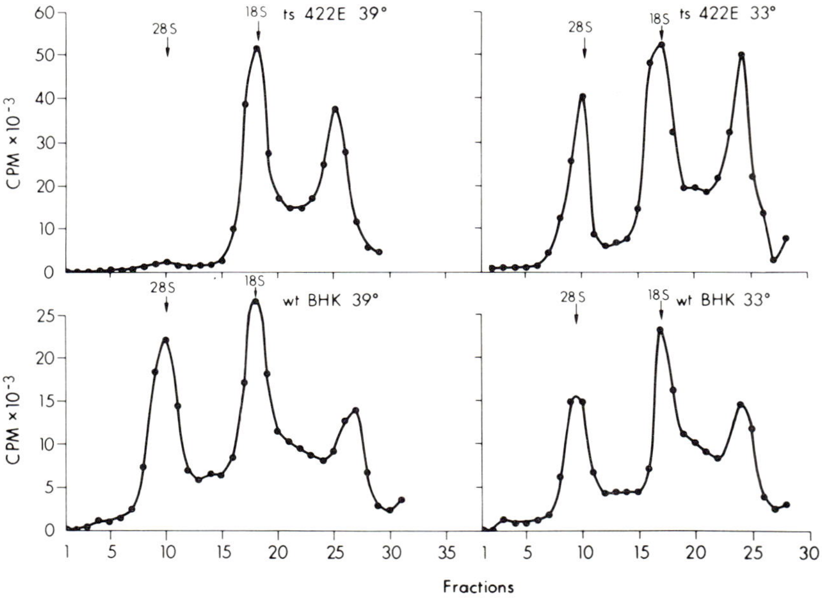

FIG. 1. Production of cytoplasmic rRNA in ts 422E and wt BHK cells at 33° and 39°. Cells were plated at 33° and 39°. After 2 days they were labeled for 2 hr at 33° and 1.5 hr at 39° with [^{3}H] uridine (1μCi, 0.2μg/ml). The cytoplasm from 1×10^7 cells was analyzed on SDS-sucrose gradients. Sedimentation is from right to left. The *arrows* indicate the position in the gradients of the A 260 nm peaks of 28S and 18S rRNA, obtained from collection of gradients through an automatically recording Gilford spectrophotometer. (From ref. 2.)

39°C. In contrast, in the mutant, there are practically no 28S rRNA subunits made at 39°C, whereas at 33°C the pattern is practically normal.

The phenomenon can be better analyzed by pulse labeling the cells for 30 min with ^{3}H uridine and then looking at the appearance of counts in 18S and 28S rRNA in the cytoplasm. The mutant produces the 18S subunit at the same rate as the wild type at 39°C. The mutant, however, incorporates the label only very slowly into the 28S peak at 39°C. It can be estimated that there is about 95% inhibition of the production of 28S subunits by the mutant at 39°C, while the synthesis of the 18S subunit is not affected (2).

Further analysis by shift-up experiments shows that upon shift of the cells to 39°C, the ability to make 28S RNA is lost rapidly. However, the temperature-sensitive product, which is made at 33°C, seems to remain functional after the shift to 39°C.

The defect in this mutant seems to be a defect in the processing of ribosomal-RNA precursors. After production of the 32S precursor, there seems to be a block, since 32S RNA accumulates in the nucleus to some extent. More important, 28S rRNA is never found in the cells, neither in the cytoplasm nor in the nucleus.

REFERENCES

1. Meiss, H., and Basilico, C. (1972): Temperature sensitive mutants of BHK 21 cells. *Nature New Biol. 239*:66.
2. Toniolo, D., Meiss, H., and Basilico, C. (1973): A temperature sensitive mutation affecting 28S ribosomal RNA production in mammalian cells. *Proc. Nat. Acad. Sci. 70*:1273.

DISCUSSION

Pontecorvo: What is the chromosome number of the mutants?

Basilico: All the mutants we have isolated have less than the diploid number of chromosomes (44). You could speculate that they are partial haploids. All the mutants have 39 to 42 chromosomes.

Siniscalco: In your series of ts mutants, you had only one showing the abnormality in rRNA synthesis.

Basilico: Yes. We have looked at four or five others that had a similar phenotype, but none of them had this same type of abnormality. In fact, this mutant seems to complement with all the others.

Siminovitch: In reference to Dr. Pontecorvo's question, in all of our mutants, the number of chromosomes has remained unchanged. We do not know if there are rearrangements and other changes which would not affect the chromosome number.

Attardi: Have you looked at 5S RNA production?

Basilico: We have not really looked thoroughly. We have looked at messenger RNA, i.e., poly-A containing RNA in the nucleus and in the cytoplasm, and that seems to be normal.

Thompson: Have you looked at ribosomal proteins?

Basilico: We are doing that now.

Littlefield: I am surprised that a cell without functional ribosomes can get so big and also that it was possible to be so logical in looking for ribosomal malfunction as you described.

Basilico: The logic was as described, and we know now that 32S RNA accumulates in the nucleoli to levels about 10- to 12-fold higher than in the wild type. Whether that is the reason why the nucleoli get bigger, I do not know. We really do not know why the cells stop dividing. It is not because they can not make proteins.

Gelehrter: Is there any defect in methylation?

Basilico: Preliminary studies suggest that there is no defect in methylation.

Somatic Cell Hybridization, edited by
Richard L. Davidson and Felix de la Cruz.
Raven Press, New York © 1974.

Regulation and Gene Expression in Heterokaryons

Nils R. Ringertz

Institute for Medical Cell Research and Genetics, Medicinska Nobelinstitutet, Karolinska Institutet, 104 01 Stockholm 60, Sweden

I. INTRODUCTION

The term *heterokaryon* is used to describe multinucleated cells formed by the fusion of two different cells. The parental cells may differ with respect to genotype or epigenotype. In interspecific heterokaryons (e.g., mouse-man) the difference is clearly genotypic. In intraspecific heterokaryons, however, the difference between the two parental cells is often relatively small as in fusions of cells from two different mouse strains or in fusions of normal and neoplastic cells from the same tissue. In some instances, the parental cells show marked and stable differences in phenotype as is the case when liver cells and lymphocytes are fused. The parental cells are then said to differ with respect to their epigenotype. In the following discussion multinucleated cells are classified as heterokaryons if there was a definable genotypic difference or stable epigenetic difference between the two parental cells. The term *homokaryon* will be reserved for polykaryocytes formed by fusion of cells among which there is no known disparity in genotype or epigenotype.

The aim of this chapter is to summarize the available information about heterokaryons with respect to the regulation of nucleic acid synthesis and phenotypic expression.

II. IDENTIFICATION OF HETEROKARYONS

If inactivated Sendai virus is added to a mixture of two cell types, a large number of heterokaryons and homokaryons are formed (24). The recognition of heterokaryons among a large number of homokaryons is often difficult. In favorable cases, such as when the parental cells differ with respect to size, shape, or chromatin structure of their nuclei, the heterokaryons can be identified on the basis of morphology. This type of identification is, however, not very reliable since the morphology of the nuclei may change after cell fusion. A more reliable method is to use ^{3}H-thymidine labeling and autoradiography. One of the parental cells is exposed to ^{3}H-thymidine for a length of time that is sufficient to allow all cells to pass one *S*-period and become labeled. The resultant heterokaryons can easily be identified in autoradiograms since they will contain both labeled *and* unlabeled nuclei within the same cytoplasm. Interspecific heterokaryons can often be identified by immune cytochemical techniques, using fluorescein labeled antibodies to cell surface or nuclear membrane antigens.

III. EARLY CHANGES AFTER FUSION

Within a few minutes after fusion of two cells carrying different surface antigens, there is a rapid intermixing of the antigens over the entire cell surface (23, 54). Shortly after cell fusion it is, however, sometimes possible to detect patches of cell surface which derive mainly from one parental cell. Thus Gordon and Cohn (15), after fusing macrophages with melanoma cells, found patches of membrane ATPase characteristic of the macrophage parent for 12 hr after fusion. This activity, however, gradually acquired a more diffuse distribution and disappeared. The cytoplasms of the cells participating in the fusion appear to mix very quickly after fusion and at the same time a redistribution of cell organelles occurs. Nuclei tend to accumulate in the center of the heterokaryon. This process can be blocked by colcemid (15), suggesting that microtubules may be important for the reorganization of intracellular architecture following fusion.

IV. FUSION OF CELLS AT DIFFERENT STAGES OF THE CELL GROWTH CYCLE

A. Homophasic and Heterophasic Heterokaryons

Fusion of two asynchronous cultures of parental cells with the aid of Sendai virus produces polykaryocytes, both heterokaryons and homokaryons, which contain different combinations of G_1, S, and G_2 nuclei. Polykaryocytes containing nuclei at more than one stage are said to be *heterophasic* whereas cells that, by chance or by design, contain nuclei at one stage of the cell growth cycle are classified as *homophasic* (Fig. 1). Stadler and Adelberg (50) have found that the fusion capacity is minimal during G_1 phase and rises to a maximum in mitosis. In most other reports there has, however, been no mention of any preferential fusion at a given stage of the cell growth cycle (29–31, 57).

B. Synchronization of DNA Synthesis

Homokaryons and heterokaryons examined some time after cell fusion, as a rule, show synchronization of DNA synthesis, i.e., either all or no nuclei synthesize DNA (29, 30, 36, 37, 55, 57). An exception to this rule is the

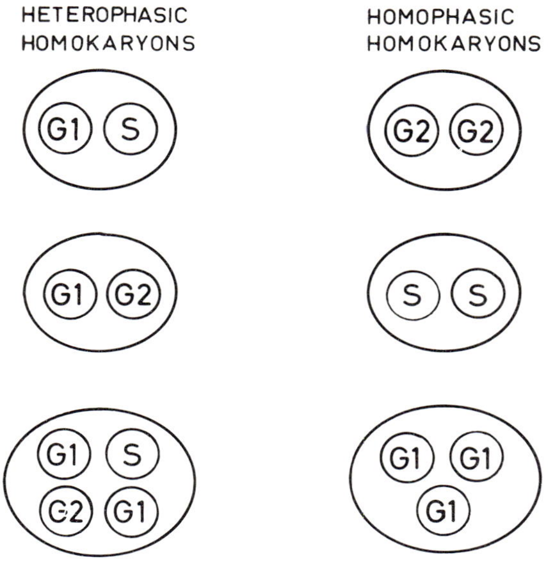

FIG. 1. Heterophasic polykaryocytes (heterokaryons or homokaryons) are formed from cells representing different stages of the DNA replication cycle. In homophasic polykaryocytes all nuclei are at the same stage with respect to DNA replication.

negative synchrony found in HeLa/Ehrlich heterokaryons studied by Johnson and Harris (31). In these heterokaryons, DNA synthesis in the HeLa nuclei was suppressed when Ehrlich nuclei were replicating. The factors immediately responsible for induction of DNA synthesis do not seem to be species or tissue specific. Thus all G_1 nuclei that share a common cytoplasm are normally forced to simultaneously initiate DNA synthesis, irrespective from which tissue or species the nuclei come. When there are exceptions to this rule, it may be because there is competition among the nuclei for a limited supply of inducer molecules (high proportion of G_1/S nuclei) or because the nucleus is unprepared for the stimulus as is the case when the interphase chromatin is excessively condensed or when the cell is in mitosis or G_2.

C. Mitosis and Premature Chromatin Condensation

The synchronization of DNA synthesis seen in polykaryocytes formed by spontaneous or induced cell fusion is often paralleled by a synchronization of mitosis (13, 29, 30, 39, 41). If, e.g., a multinucleated cell contains several G_1 nuclei (homophasic polykaryocyte), the different G_1 nuclei will initiate DNA synthesis simultaneously and also enter mitosis in synchrony.

Fusion of a mitotic cell with another in interphase results in a precocious attempt of the interphase nucleus to enter mitosis. Chromosome-like structures form in the interphase nucleus and the nuclear membrane dissolves. This phenomenon has been termed *premature chromatin condensation* (PCC) by Johnson and Rao (32, 33) and is probably identical to the chromosome pulverization observed in cultures infected with measles virus (40) and other fusing viruses (35). Through the work of Stenman and Saksela (51) and Johnson and Rao (32), it became clear that the phenomenon previously known as chromosome pulverization results from the fusion of an interphase cell with a cell in mitosis and represents a premature and incomplete condensation of interphase chromosomes. In view of this, the term premature chromatin condensation is more appropriate than chromosome pulverization, a term that has also been used to describe other forms of chromosome aberrations.

The PCC phenomenon has been analyzed by Johnson and Rao (32, 33) in experiments with synchronized HeLa cells and UV-inactivated Sendai virus. One population of synchronized cells was fused with another at the same time or at a different stage in the cell growth cycle. In multinucleated cells containing both mitotic and interphase nuclei, the latter were found to undergo PCC. G_1 nuclei produced chromosomes that were considerably thinner than normal mitotic chromosomes whereas G_2 nuclei produced chromosomes that were more similar to those present during normal mitosis. Furthermore, the chromosome filament derived from G_1 nuclei appeared to consist of a single chromatid, whereas G_2 nuclei produced double chromatids (Fig. 2). *S* phase nuclei only condensed partially and gave rise to irregular fragmented chromatin masses. By analyzing cells containing varying proportions of mitotic and interphase nuclei, Johnson and Rao (32) found that PCC was far more readily induced in G_1 nuclei than in either *S* or G_2 nuclei. Furthermore there was a clear dose dependence between the induction of PCC and the proportion of mitotic to interphase nuclei; PCC being induced much more rapidly if the ratio of mitotic to interphase nuclei was high. The signals that induce PCC in fusions of mitotic and interphase cells do *not* appear to be species specific since PCC is observed in interspecific heterokaryons (27, 32).

The factors inducing PCC appear to be present only in mitotic cells since PCC is never observed in heterophasic polykaryocytes containing G_1, *S*, and/or G_2 nuclei. Matsui et al. (38) have demonstrated that the factor necessary for PCC is synthesized 15 to 45 min before the onset of normal metaphase. The critical factor is not formed if protein synthesis is blocked during the G_2 phase. The chemical nature of the PCC inducing protein(s) has not yet been elucidated and it is possible that several factors are involved. Matsui et al. (38) discussed the possibility that there may be different factors involved, some of which cause the condensation, whereas others produce the chromosome fragmentation seen when PCC is induced in

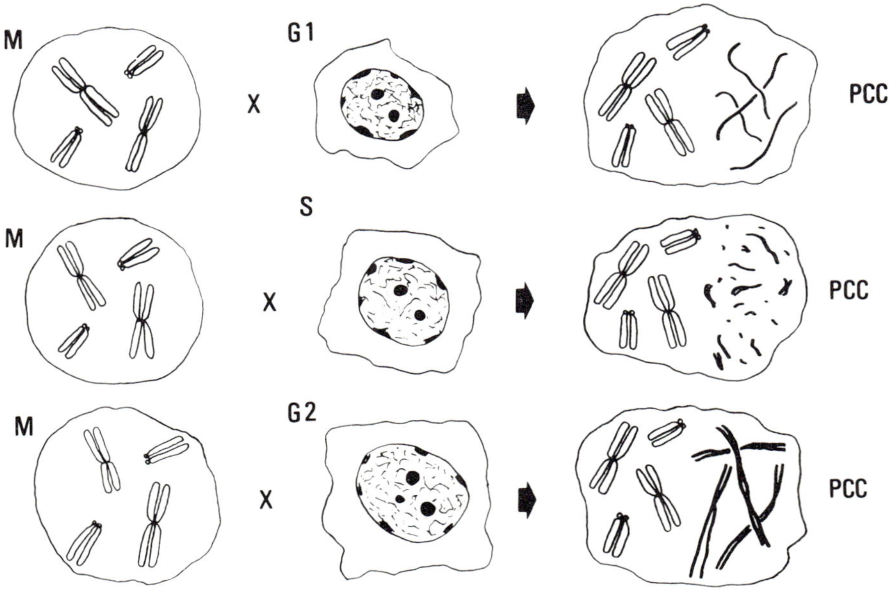

FIG. 2. Schematic illustration of the appearance of prematurely condensed chromosomes (PCC) in fusions of mitotic *(M)* cells with interphase cells in G_1, *S*, or G_2. In $M \times G_1$ fusions the PCC filaments correspond to single chromatids while in $M \times G_2$ fusions two chromatids are present. $M \times S$ fusions result in condensation and fragmentation of the chromatin.

S nuclei. The latter factor may be nucleases, which are known to be particularly active in *S* cells. The reason why PCC is induced most easily at the G_1 stage and the basis for the dosage effect (i.e., that PCC is induced most rapidly if the ratio of mitotic to interphase nuclei is high) cannot yet be fully explained. One possible explanation may be that the susceptibility of the deoxyribonucleoprotein to the condensing effect of inducers varies during the cell growth cycle and that greater quantities of mitotic inducers are needed to condense more or larger (G_2) interphase nuclei. Further studies of the PCC phenomenon are clearly needed. The fact that large populations of synchronized cells can be fused with one another and the high efficiency of the PCC induction has undoubtedly presented us with new possibilities of analyzing mitosis by biochemical methods. The possibility of inducing PCC is also of considerable interest as a technique for visualizing chromosomes in interphase cells that do not normally divide. Thus premature chromosome condensation has been induced in unstimulated horse lymphocytes, bovine spermatozoa, Chinese hamster ovary cells, embryonic chick fibroblasts and erythrocytes, *Xenopus* kidney and mosquito cells by fusion with colcemid blocked HeLa cells (34).

V. REACTIVATION OF DORMANT CELL NUCLEI IN HETEROKARYONS

When Harris and Watkins (24) and Harris et al. (25, 26) introduced the use of Sendai virus as a means of producing multinucleated hybrid cells, they noted that inactive cell nuclei, among them chick erythrocyte nuclei and mouse macrophage nuclei, underwent a marked reactivation and showed an accelerated synthesis of RNA. The reactivation of chick erythrocyte nuclei has then been further studied by Harris (20) and by our group in Stockholm (for a review, see ref. 45). The reactivation of chick erythrocyte nuclei in heterokaryons has provided an interesting system for the analysis of nucleocytoplasmic interaction and its role in the control of nuclear activity. Macrophage heterokaryons, on the other hand, have provided new information about factors controlling phenotypic expression (15–18). The macrophage heterokaryons, therefore, will be dealt with mainly in connection with phenotypic expression in heterokaryons.

A. Chick Erythrocyte Heterokaryons

Until quite recently, it was generally believed that the inactivation of the erythrocyte nuclei during erythropoiesis was a more or less irreversible process. This situation changed when Harris (19, 20) demonstrated that the chick erythrocyte nucleus is reactivated in heterokaryons formed by Sendai virus-induced cell fusion of chick erythrocytes with fibroblasts, human tumor cells (Fig. 3) or other active cells. The reactivation process involved a

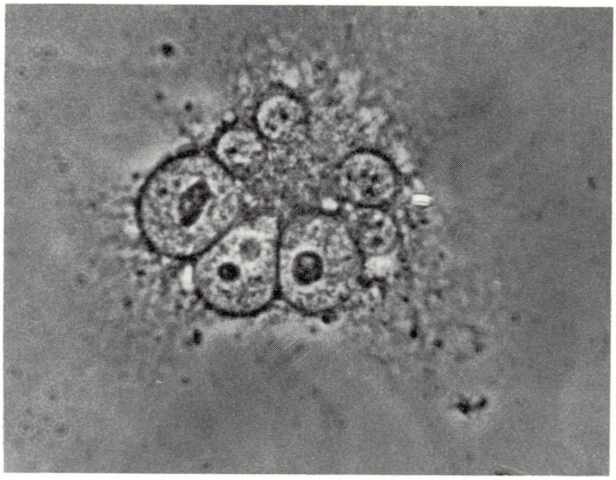

FIG. 3. Heterokaryon formed from a fusion of 3 human tumor cells (HeLa) with 4 chick erythrocytes (smaller nuclei). (From ref. 46.)

series of morphological and chemical changes. The main events during reactivation in heterokaryons made with human tumor cells (HeLa) can be summarized as follows:

(1) early changes in the physicochemical properties of deoxyribonucleoprotein ("chromatin activation") (1, 43, 44),
(2) increase in nuclear volume and dry mass and a dispersion of the condensed chromatin (1, 20),
(3) migration of *human* nucleospecific proteins into the chick nucleus (10, 46),
(4) initiation of HnRNA synthesis (high-molecular weight, polydisperse RNA (19, 23),
(5) migration of human nucleolo-specific proteins into the chick nucleus (10, 46), and the formation of a nucleolus (20, 49),
(6) initiation of ribosomal RNA (rRNA) synthesis (23),
(7) initiation of chick specific protein synthesis including chick nucleolar antigens, surface antigens, enzymes, etc. (7, 22, 23, 46),
(8) migration of chick nucleolar components into the chick nucleolus (46), as well as into the nucleoli of the HeLa nuclei,
(9) DNA synthesis (1, 19),
(10) formation of mononucleate synkaryons (46, 48).

1. "Chromatin Activation"

At an early stage of the reactivation of the chick erythrocyte nucleus in HeLa cells, the DNA-protein complex (DNP) undergoes drastic physicochemical changes that manifest themselves in an increased binding of basic dyes, an increased susceptibility to thermal denaturation, and acid hydrolysis (1). These changes may be due to a removal of proteins and/or divalent metal ions in the DNP complex as well as to more complex conformational changes. Similar nucleoprotein changes can be induced in chick erythrocyte ghosts or nuclei by manipulating the ionic environment (44). The first signal in the reactivation of the chick erythrocyte nucleus may therefore be changes in its ionic environment caused by contact with HeLa cytoplasm. For a more detailed discussion of this phenomenon, see refs. 43 and 44.

2. Nuclear Growth

The increase in nuclear volume during the reactivation of the chick erythrocyte nucleus is paralleled by an increase in nuclear dry mass. Most of the dry mass increase appears to be due to an accumulation of protein in the erythrocyte nuclei. Cytochemical (2), immunological (46), and biochemical (Goto and Ringertz, *to be published*) data suggest that a considerable proportion of the proteins accumulating in the erythrocyte nucleus are human proteins. The migration of human macromolecules into the chick erythrocyte nucleus during the reactivation process has been studied with the aid of the indirect immune fluorescence and antibodies to human nuclear proteins collected from patients with various "autoimmune" diseases. Human autoantibodies bind to antigens in human cells including fixed HeLa cells, but not to antigens in chick cells. At early stages of reactivation chick erythrocyte nuclei in heterokaryons do not bind antibodies directed against human nuclei.

As the nuclei enlarge, however, human antigens can be detected in the nucleoplasm, nuclear membrane, and nucleolus of the chick erythrocyte nucleus. Only antigens characteristic of human cell nuclei accumulate in the growing erythrocyte nucleus. Antigens characteristic of human cytoplasms do not enter the chick nuclei. Therefore the growth of the chick erythrocyte nucleus in heterokaryons made with HeLa cells is not due to a random uptake of material from the surrounding cytoplasm but to a selective concentration of human nucleospecific macromolecules. The individual human antigens show exactly the same distribution in the fully reactivated erythrocyte nucleus as they do in the HeLa nucleus. Human nuclear membrane, nucleoplasmic and nucleolar antigens are thus found in equivalent structures in the chick nuclei and are not randomly distributed in the chick nucleus. This phenomenon may be due to selective penetration of antigenic macromolecules through the nuclear membrane and selective binding in the nuclear compartment. Altered properties of the DNP, such as those observed in this system, may be the reason for the trapping of certain proteins (antigens) in the chick nucleus. The "activated" DNP molecules may function as an ion exchange resin and retain proteins in the nuclear compartment. The characteristic intranuclear distribution of human nucleolar and nucleoplasmic antigens in the chick nucleus is probably due to a specific affinity for specialized DNP regions. The fact that human membrane antigens are concentrated in the chick nuclear membrane could be due to an affinity of membrane protomers for preexisting membrane structures, irrespective of species differences.

The chemical nature of the proteins migrating into the chick erythrocyte nucleus from the cytoplasm is not yet known. Using a cytochemical method for endogenous RNA polymerase, Carlsson et al. (5) found that both nucleoplasmic and nucleolar types of RNA-polymerase activity increased during the activation of chick erythrocyte nuclei in rat epithelial cells. Data obtained by Darzynkiewicz and Chelmicka-Szorc (8) indicate that DNA–repair enzymes may be among the macromolecules that migrate into the chick nucleus. We have also tried to examine if HeLa cell RNA migrates into the chick erythrocyte nuclei during their reactivation in heterokaryons. A migration of certain types of RNA from the cytoplasm into the nucleus has previously been demonstrated in nuclear transplantation experiments with amoeba (14). No such migration could be detected in HeLa–chick erythrocyte heterokaryons (Ege and Ringertz, *unpublished observations*). In relative terms human RNA appears to be selectively excluded from entering the growing chick nucleus. Our results do not exclude, however, the possibility that small quantities of human RNA do migrate into the chick nucleus.

3. *Nucleic Acid Synthesis*

Shortly after fusion of chick erythrocytes with HeLa cells, the erythrocyte nucleus resumes RNA synthesis (20). As the erythrocyte nucleus grows in size and the chromatin becomes more dispersed, the rate of labeling with ^{3}H-uridine increases. Autoradiographic data obtained by Harris (20) and confirmed by Carlsson et al. (4), suggest that the rate of RNA synthesis increases in direct relation to nuclear growth. The RNA produced in the erythrocyte nucleus during the first days after fusion has a high molecular weight and is polydisperse (23). This RNA is not transported to the cytoplasm and appears to be similar to the HnRNA found in other cell systems. Ribosomal RNA (rRNA) cannot be detected until about the third day after fusion. The appearance of these RNA fractions coincides with the appearance of nucleoli visible under the light microscope. Transport to the cytoplasm of RNA synthesized in the chick nucleus and new synthesis of chick proteins (surface antigens) can only be detected from the stage where nucleoli develop (49) and rRNA is synthesized. The continued synthesis of chick proteins in heterokaryons containing a fully reactivated erythrocyte nucleus also appears to be dependent upon nucleolar function. If the only nucleolus in a single erythrocyte nucleus is irradiated with a UV-microbeam, the synthesis of chick proteins ceases even if nucleoplasmic RNA synthesis continues in the erythrocyte nucleus (9). The late activation of nucleolar RNA synthesis observed by Harris and co-workers is paralleled by a late appearance of nucleolar RNA polymerase activity (5).

After some 10 to 15 hr in the hybrid, the majority of the erythrocyte nuclei have increased markedly in volume and dry mass but still contain the G_1 amount of DNA. Some of the chick erythrocyte nuclei have, however, begun to incorporate thymidine as originally shown by Harris (19). Some 20 hr after hybridization many erythrocyte nuclei have higher Feulgen-DNA values than control erythrocytes and ultimately a majority of the nuclei can be found to have the premitotic amount of DNA (1). This clearly shows that the incorporation of thymidine reflects a net synthesis of DNA and furthermore that the erythrocyte nuclei are capable of replicating their DNA completely.

There is no close correlation between the stage of nuclear growth and the initiation of DNA synthesis. Some nuclei may show at least a threefold increase in dry mass without any increase in DNA amount, whereas others may have enlarged very little but still be in the process of DNA replication. Some enlargement seems to be necessary, however, before DNA synthesis can be initiated in the erythrocyte nuclei.

B. Macrophage Heterokaryons

Macrophages can be obtained from the peritoneal fluid of rabbits, mice, or rats and can be purified from other cells by virtue of their ability to attach rapidly and firmly to glass surfaces. Macrophages do not normally synthesize DNA but appear to be arrested in the G_0 phase in the same way as many other highly differentiated cells. Their kidney-shaped cell nuclei do, however, synthesize RNA and have small nucleoli. The nucleus is smaller than that of more active cell nuclei from different tissues but of the same species. When macrophages are fused with more active cells such as HeLa cells (26) and melanoma cells (15) the macrophage nucleus becomes swollen, the nucleoli become more prominent, and RNA synthesis accelerates by a factor of from 4 to 10. The macrophage nuclei are also triggered to resume DNA synthesis and some heterokaryons also enter mitosis. These phenomena do not occur if the macrophage fuses with other macrophages (homokaryons) or with other inactive cells, e.g., lymphocytes.

The reactivation of macrophage nuclei in macrophage-melanocyte heterokaryons appears to be a more rapid process than the reactivation of chick erythrocyte nuclei in heterokaryons made with HeLa tumor cells or melanoma cells (16). Macrophage nuclei swell and show increased RNA synthesis within an hour, and a wave of DNA synthesis starts 3 hr after fusion.

Using inhibitors of RNA and protein synthesis, Gordon and Cohn (16, 17) have examined if the reactivation of the macrophage nuclei is dependent upon RNA or protein synthesis specified by the macrophage and melanoma nuclei, respectively. Their results suggest that the reactivation of DNA synthesis in the macrophage nuclei is dependent upon RNA synthesized by the *melanoma* nucleus. Inhibition of RNA synthesis in the macrophage nucleus, on the other hand, does not prevent the initiation of macrophage-DNA synthesis. Pretreatment of the melanoma parent with inhibitors of protein synthesis before fusion, effectively prevents the initiation of DNA synthesis in the macrophage nuclei after fusion. Pretreatment of the macrophage parent before fusion, however, does not prevent the initiation of macrophage-DNA synthesis. These results suggest that the melanoma cells provide the macrophage nuclei with products that trigger and make possible their DNA synthesis. Also in the macrophage heterokaryons formed with *S*-phase melanoma cells, a preliminary swelling of the inactive nucleus appears to be required before DNA synthesis is initiated. The lag period that precedes the reactivation of DNA synthesis in macrophage nuclei is, however, shorter than that in the erythrocyte nuclei. The explanation for the more rapid reactivation of both RNA and DNA synthesis in macrophage nuclei may be that from the start, these nuclei are less condensed and less completely inactivated than erythrocyte nuclei.

VI. REGULATION OF NUCLEIC ACID SYNTHESIS IN HETEROKARYONS

Table 1 represents an attempt to summarize results obtained with different types of heterokaryons. Three major conclusions can be drawn from this work. One is that the active partner in the fusion always seems to *stimulate* the nucleus of the inactive partner to synthesize more RNA and/or DNA. So far there is no indication that the inactive parental cell would inhibit or repress the more active nucleus. The second conclusion is that the inactive nucleus seems to be activated to about the same level as the active nucleus

TABLE 1. *RNA and DNA synthesis in some heterokaryons and their parental cells*

PARENTAL CELL TYPE	RNA	DNA
HeLa (human)	+	+
A9 "fibroblast" (mouse)	+	+
Established fibroblast (monkey)	+	+
Melanocyte (mouse)	+	+
Myoblast (rat)	+	+
Myotube (rat)	+	−
Macrophage (rabbit)	+	−
Lymphocyte (rat)	+	−
Neuron (mouse)	+	−
Erythrocyte (chick)	−	−

HETEROKARYON Cell I	Cell II	RNA synthesis Nucleus I	RNA synthesis Nucleus II	DNA synthesis Nucleus I	DNA synthesis Nucleus II	References
HeLa	Macrophage	+	*	+	*	19
HeLa	Lymphocyte	+	*	+	*	19
HeLa	Erythrocyte	+	*	+	*	19, 26
A9	Erythrocyte	+	*	+	*	23
Macrophage	Lymphocyte	+	+	−	−	26
Macrophage	Erythrocyte	+	*	−	−	26
Myoblast	Erythrocyte	+	*	+	*	6
Myotube	Erythrocyte	+	*	−	−	6
Melanocyte	Macrophage	+	*	+	*	16, 17, 18
Fibroblast	Neuron		?	+	+	28

Explanation of symbols:
\+ synthesis
− no synthesis
* induced or accelerated synthesis
? not tested

but not beyond this level. Thus where the active cell is synthesizing RNA but not DNA (macrophages, myotubes), the inactive nucleus is only stimulated to make RNA. If, on the other hand, the active partner is synthesizing both types of nucleic acid, then the inactive nucleus is also stimulated to synthesize both types of nucleic acid. The third conclusion is that the signals that regulate the overall level of RNA and DNA synthesis do not seem to be species specific. Active cells of mouse origin can stimulate inactive human, rabbit, or chick as well as mouse nuclei.

The chemical nature of the signals that regulate nucleic-acid synthesis in heterokaryons is not known. As indicated above, the early nucleoprotein changes during the reactivation of chick erythrocyte nuclei in heterokaryons can be mimicked in isolated nuclei by merely manipulating the ionic environment (44). It is therefore possible that changes in the ionic environment may function as signals during the early part of the reactivation process. The rapid growth which begins when a dormant cell nucleus is reactivated appears to be due to the migration of preformed protein from the cytoplasmic compartment into the nucleus. This process appears to be passive in the sense that it is not dependent on the RNA synthesized by the nucleus undergoing reactivation (2). It may therefore be appropriate to refer to the early nuclear growth as a swelling process where macromolecules are taken up from the cytoplasm. This process may provide the previously inactive cell nucleus with factors that modify the template properties of the chromatin and also provide enzymes necessary for the transcription process (e.g., RNA polymerase). It is interesting in this connection that the macromolecules taken up from the cytoplasm by the swelling erythrocyte nucleus do not seem to represent a cross-section of the cytoplasmic macromolecules but rather a selective accumulation of macromolecules normally found in high concentrations in active cell nuclei. The nucleo-cytoplasmic protein migration also shows the lack of species specificity (46) that is necessary to explain the data summarized in Table 1.

VII. PHENOTYPIC EXPRESSION IN HETEROKARYONS

A. Markers and Patterns of Expression

The markers that have been studied in hybrid cells can be divided into those common to all cell types ("household" molecules) and those found exclusively in cells from a certain specialized tissue (histotypic markers or "luxury" molecules) (12).

The analysis of phenotypic expression in heterokaryons depends, to a great extent, on the availability of specific cytochemical or immunological reactions by which well-defined marker properties can be detected in single cells. Enzyme cytochemical methods and immune fluorescence reactions may offer great specificity but can be very difficult to quantitate. Autoradiography, after supplying the cells with radioactive precursors, can sometimes be used to detect the presence or absence of biochemical pathways characteristic of a certain type of cell. From a methodological point of view, however, populations of mononucleated hybrid cells offer much better possibilities for a detailed analysis of phenotypic expression, since large numbers of cells can be obtained and then analyzed by standard biochemical method. There are, however, also some problems with this type of analysis. Reasonably homogeneous populations of mononucleated hybrid cells can, as a rule, only be obtained after the cells have spent a relatively long time in culture. During this time there is a marked selection for those hybrid cells that are capable of rapid multiplication. Furthermore, the standard tissue-culture procedure selects hybrid cells that grow well on plastic and that are well adapted to the presence of sera from foreign animals, synthetic buffer substances, pH-indicators, etc. During prolonged culture, interspecific hybrids lose chromosomes and undergo chromosomal rearrangements. This made chromosome mapping possible but the chromosomal instability of the hybrids may, at the same time, be a problem in the study of regulatory processes. Thus the absence of a certain marker in the phenotype of the hybrid cell can be due to a regulatory process but equally well to a loss of the chromosome carrying the structural gene that specifies the marker. Only in those cases where, after some time, the marker is again reexpressed can one be sure that the phenomenon studied was indeed regulatory.

The advantage with heterokaryons is that phenotypic expression can be studied from the time of fusion and before any selection or chromosome losses have occurred. It is often possible to examine the role of gene dosage in phenotypic expression since virus-induced fusion results in heterokaryons containing varying numbers and proportions of the parental nuclei. Since single cells are analyzed, one can use culture conditions that favor cell differentiation even if this means that cell multiplication is inhibited.

Although so far only relatively few studies have been made of phenotypic expression in heterokaryons (Table 2), different expression patterns have been recognized. Most hybrid cells whether heterokaryons or mononucleated hybrid cells express properties characteristic of both parental cells (*co-expression* or *codominance*). Thus Watkins and Grace (54) have found surface antigens characteristic of both parental cells in heterokaryons made by fusing human HeLa cells with Ehrlich cells. A similar case of coexpression was observed by Carlsson et al. (4) in studies of hybrid myotubes. Myotubes made by fusing chick and rat myogenic cells synthesized both chick and rat myosin. In other types of heterokaryons it has been observed, however, that properties characteristic of one parental cell are *extinguished* and cannot be detected in the heterokaryons. Extinction of differentiated properties has been observed by Gordon and Cohn (15, 18) in macrophage melanoma heterokaryons, where the macrophage marker properties disappear. In heterokaryons formed by fusing rat hepatoma cells with rat

TABLE 2. *Phenotypic expression in heterokaryons*

Coexpression marker	Parental cells	References
antigens, mouse + human	Ehrlich (mouse)–HeLa (human)	54
myosin, chick + rat	myoblast (chick)–myoblast (rat)	4
hemoglobin, frog + tadpole	erythroblast (frog)–erythroblast (tadpole)	47
Extinction marker	**Parental cells**	**References**
phagocytosis	macrophage (mouse)–melanoma (mouse)	15–18
lysosomal enzymes	macrophage (mouse)–melanoma (mouse)	15–18
membrane ATP: ase	macrophage (mouse)–melanoma (mouse)	15–18
hemoglobin	erythroblast (chick)–A9 (mouse)	21
Inducibility of tyrosine amino transferase (TAT)	hepatoma (rat)–epithelial cells (mouse)	53

epithelial cells, Thompson and Gelehrter (53) observed that the inducibility of tyrosine aminotransferase (TAT), a property characteristic of liver differentiation, disappeared in the heterokaryon. These examples illustrate the two most important patterns of phenotypic expression, *coexpression* and *extinction* of properties characteristic of the parental cells. Two types of heterokaryons have been examined in much greater detail than other heterokaryons: chick erythrocyte and macrophage heterokaryons. Since the pattern of phenotypic expression is better known in these heterokaryons we shall discuss these two types of heterokaryons in more detail.

B. Erythrocyte Heterokaryons

Heterokaryons, containing fully reactivated chick erythrocyte nuclei with well developed nucleoli, synthesize many different chick macromolecules. The spectrum of chick proteins produced in HeLa or mouse L cell heterokaryons includes surface antigens (23), at least two different enzymes involved in nucleotide synthesis, hypoxanthine guanine phosphoribosyl transferase and adenosine phosphoribosyl transferase (22) (also Harris, *personal communication*), and nucleolus specific antigens (46). There has not, however, been any indication that heterokaryons made by fusion of erythrocytes with undifferentiated cells produce hemoglobin or other marker molecules typical of erythroid differentiation (21). If immature erythroid chick cells that are synthesizing chick hemoglobin are fused with undifferentiated mouse cells, the hemoglobin synthesis stops within a relatively short time after fusion. Heterokaryons produced by fusing tadpole and frog erythrocytes synthesize both tadpole and frog hemoglobins. In comparison with homokaryons frog hemoglobin synthesis is greatly stimulated while tadpole hemoglobin synthesis is depressed (47). It is, however, still quite difficult to interpret these data in terms of regulatory events. One reason is that it is often very difficult to get tissue-specific differentiation of normal nonhybrid cells *in vitro*. Bone-marrow culture containing erythroid precursor cells requires the addition of erythropoietin to undergo differentiation. It can be expected that the heterokaryons are no less demanding and therefore that the phenotype of erythrocyte heterokaryons should be examined under conditions where nonhybrid cells are capable of differentiation.

By reactivating chick erythrocyte nuclei in the cytoplasms of rat and mouse cells representing different lines of cell differentiation, we have attempted to study how the cytoplasm influences the phenotype expressed by the reactivated chick erythrocyte nucleus (4). More specifically we have been interested to see if it is possible to reprogram the chick erythrocyte nucleus to specify a phenotype similar to that of the other parental nucleus. Chick erythrocyte nuclei have therefore been reactivated in rat myoblasts and rat myotubes at a time when rat myosin synthesis starts (6). No significant chick myosin synthesis could be demonstrated in spite of the fact that the chick erythrocyte nuclei developed a nucleolus, produced RNA, and expressed other chick genes in the form of chick antigens. At the same time heterokaryons made by fusing chick myoblasts with rat myoblasts synthesized both chick and rat myosin (4).

C. Macrophage Heterokaryons

Macrophages possess several well-defined marker properties that make them suitable as parental cells in cell fusion experiments aiming to study mechanisms regulating phenotypic expression. Gordon and Cohn (15–18) have studied macrophage heterokaryons with respect to the expression of the following three markers of macrophage differentiation:

(1) the ability to bind and phagocytose antibody coated red cells, dextran sulphate and other macromolecules,
(2) the presence of a Mg^{++} activated membrane bound ATPase,
(3) the presence of lysosomes and lysosomal enzymes (acid phosphatase).

Heterokaryons formed by fusing macrophages with undifferentiated melanoma cells lose the capacity to phagocytose after a few days. The rate at which the phagocytosing capacity is lost is dependent upon the proportion of macrophage to melanoma nuclei in the individual heterokaryon. A higher proportion of "undifferentiated" melanoma nuclei results in a more rapid inactivation of phagocytosis. Initially the macrophages have about 30 times more surface ATPase activity than the melanocyte. After fusion the ATPase activity is lost in heterokaryons at a rate that is dependent upon the ratio macrophage/melanocyte nuclei whereas at the same time homokaryons containing only macrophage nuclei retain their ATPase activity.

Other macrophage markers such as lysosomes and lipid droplets also disappear from the heterokaryons whereas they are retained by the macrophage homokaryons. The disappearance of phagocytosis in macrophage heterokaryons appears to be due to a masking of receptors on the cell surface. Macrophages normally carry membrane receptors that bind antibody coated particles and that can be demonstrated by mixing macrophages with sheep red cells coated with 7S antibody. Such receptors, present on the heterokaryons immediately after fusion, disappear after a short time (15). The receptors can, however, be exposed by treating the heterokaryons with trypsin. If subjected to further cultivation in the absence of trypsin, the receptors will again become masked. Inactivation of the melanoma nucleus before fusion will preserve macrophage receptors on the surface of the heterokaryons whereas inactivation of the macrophage nucleus is without effect.

The extinction of phagocytic function in macrophage heterokaryons is clearly related to the state of activity of the nonmacrophage parent (18). If this cell is a rapidly growing Ehrlich tumor cell the receptors disappear much faster than if the other parental cell is a normal primary chick fibroblast. In macrophage–chick erythrocyte heterokaryons the macrophage receptors are maintained for at least 7 days. This suggests that extinction of the macrophage markers only occurs when the nonmacrophage parent is an active cell.

VIII. RECONSTITUTED CELLS

Techniques for mass enucleation of monolayers of somatic cells adhering to glass or plastic surfaces have recently been developed (42, 56). With the aid of Sendai virus–induced cell fusion, it has also been possible to reintroduce nuclei into the enucleated cytoplasms so as to produce a functioning cell (11). In the following, I shall refer to such cells as *reconstituted cells.*

The enucleation is carried out by treating cells attached to a plastic slide with the drug cytochalasin B, at the same time as the cells are centrifuged. The cytochalasin is believed to affect microfilament systems in the cytoplasm and thus induces the formation of cytoplasmic protrusions. With the aid of the centrifugal force (Fig. 4) during centrifugation, the nucleus can be drawn out into a long cytoplasmic stalk. This stalk often breaks leaving the main part of the cytoplasm attached to the slide while the nuclei sediment to the bottom of the centrifuge tube. The cytoplasms can be recovered by taking out the slides and washing away the cytochalasin-containing medium. Such cytoplasms synthesize protein for up to 3 days but then round up and die. The enucleation can be made almost complete with certain types of cells (Fig. 5).

The nuclei, "drawn out" by the cytochalasin-centrifugation procedure, have the appearance of normal nuclei in intact cells and are surrounded by a narrow rim of cytoplasm and a plasma membrane (Fig. 6). In order to obtain large numbers of nuclei, the enucleation can be carried out by centrifuging the cells on a serum albumin density gradient containing cytochalasin. We have tried to use the enucleated cytoplasm and the cytochalasin nuclei

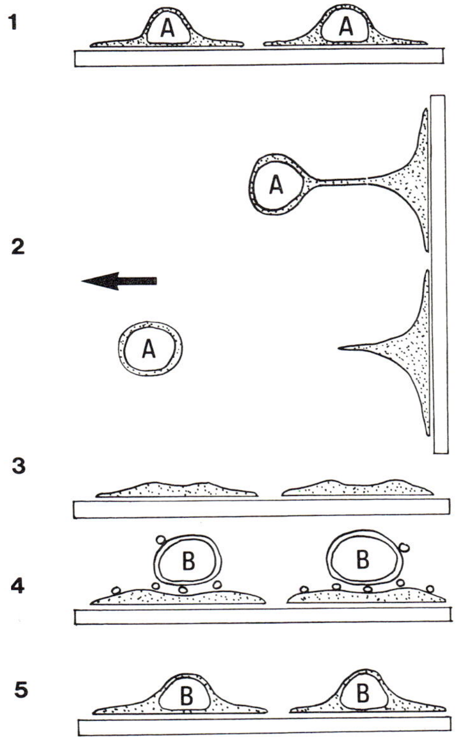

FIG. 4. Enucleation by cytochalasin treatment and centrifugation (1, 2). The enucleated cells (3) are then fused by means of Sendai virus with isolated nuclei or nucleated erythrocytes (4, 5). Experimental conditions: Cells growing in monolayer on glass slides were placed upside down in centrifuge tubes in specially constructed holders. Centrifugation was carried out at 3000 *g* for 40 min (37°C) in phosphate buffered saline containing cytochalasin B at a concentration of 10 μg/ml. (From ref. 11.)

for different types of fusion experiments aimed at studying phenotypic stability. Chick erythrocytes have been fused with enucleated L-cell cytoplasm (Fig. 7) to examine whether or not the erythrocyte nuclei would be reactivated in the absence of a "host" nucleus, i.e., in an enucleated cytoplasm. We have found that reactivation occurs at approximately the same rate as in heterokaryons. Although this type of "reconstituted" cell synthesizes RNA and expresses new chick antigens, it does not seem to divide or to survive for any longer period of time. It seems that reactivation may not be sufficiently rapid to rescue the cytoplasms from death. We are, therefore, trying other combinations of nuclei and cytoplasms.

At this stage, I would just like to point out that the possibility to glue a piece of "undifferentiated" cytoplasm to a differentiated cell or to reconstitute cells from nuclei and cytoplasms of different genotypes and phenotypes

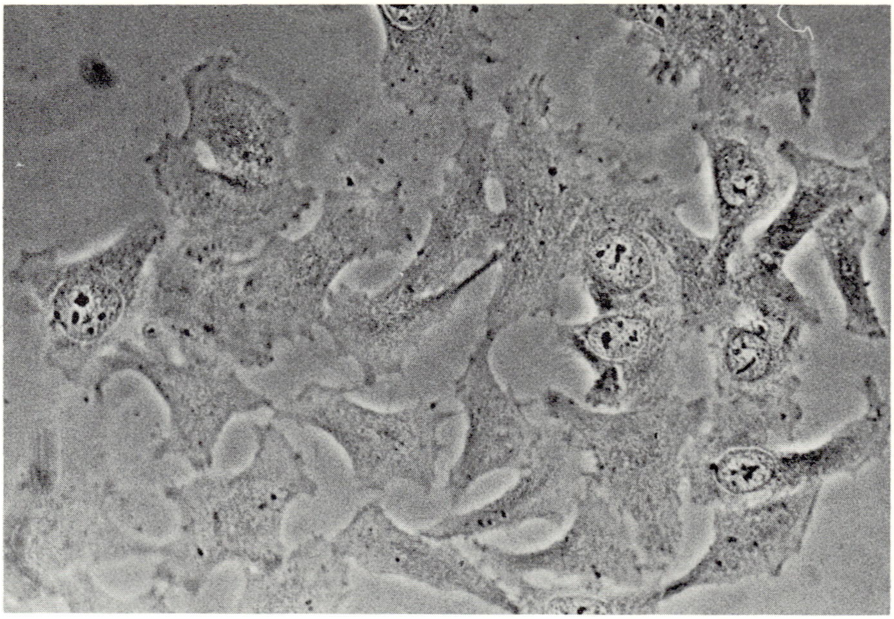

FIG. 5. Phase contrast microphotograph of enucleated HeLa cells together with a few nucleated cells. (From ref. 11.)

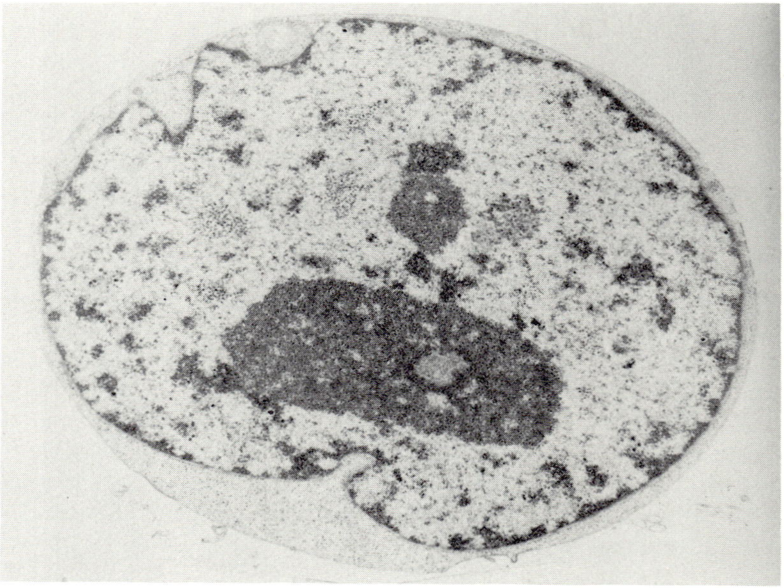

FIG. 6. Electron micrograph showing degree of cytoplasmic contamination around nuclei extracted by centrifugation from cytochalasin treated mouse L-cells (Dalen, Ege and Ringertz, *to be published*).

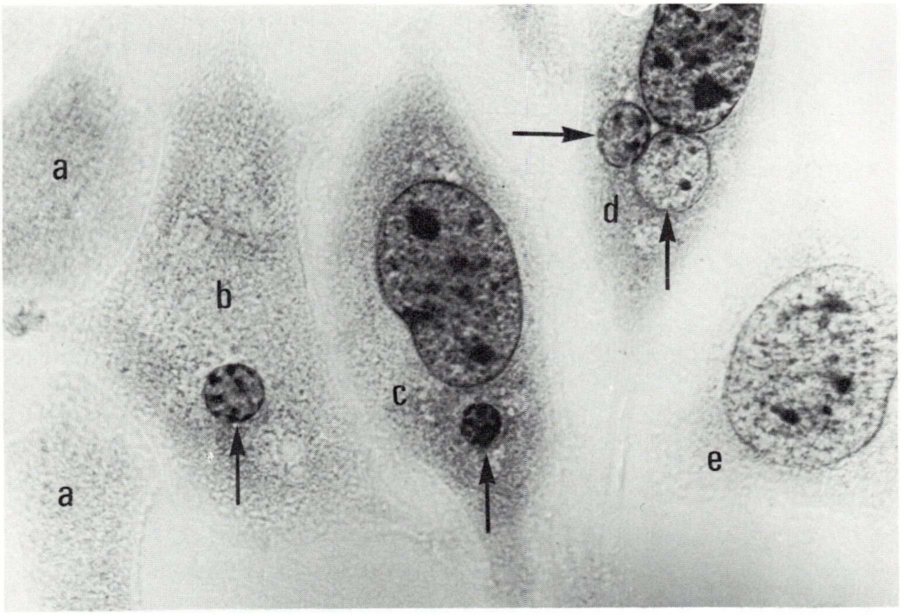

FIG. 7. Chick erythrocyte nuclei introduced into enucleated and nucleated A9 mouse fibroblasts: enucleated mouse fibroblasts (a), enucleated mouse fibroblast containing a chick erythrocyte nucleus (b), mouse fibroblast × chick erythrocyte heterokaryon with one (c) or two (d) erythrocyte nuclei, normal mouse fibroblast (e). (From ref. 11.)

offers fascinating new possibilities to analyze cell differentiation and gene regulatory mechanisms.

IX. CONCLUDING REMARKS

Cytological and cytochemical analysis of heterokaryons offers new possibilities of studying factors regulating cell division, nucleic acid synthesis and gene expression in eukaryotic cells. There are also indications that heterokaryons can provide new information about the interaction between tumor viruses and host cells (52). Furthermore it has been demonstrated that heterokaryons can be used as a system for analyzing gene complementation (3).

The biological information obtained with heterokaryons stresses the role of the cytoplasm in the control of nuclear activity. When a G_1 nucleus is brought into contact with the cytoplasm of an S-phase cell, the G_1 nucleus is stimulated to synthesize DNA. If the G_1 nucleus is combined with a mitotic cell, the chromatin of the G_1 nucleus is forced to condense into premature chromosome-like structures. Inactive nuclei, brought into contact with an active cell or enucleated cytoplasm from such cells, will be activated with respect to nucleic acid synthesis. The degree of activation of the inactive nucleus appears to be determined by the state of the active cell. Inactive nuclei are activated to the same level as the active nucleus but never beyond this level. It should also be noted that the inactive parental cell in a fusion does not inhibit the RNA or DNA synthesis of nuclei derived from an "active" parental cell.

The analysis of phenotypic expression in heterokaryons has only just begun. Examples of coexpression and extinction of histotypic markers have been reported. Extinction of histotypic markers occurs in heterokaryons where one of the parental cells is an actively growing, "undifferentiated" cell, and the differentiated cell has an inactive nucleus (macrophages, nucleated erythrocytes). Coexpression of histotypic markers has been found in heterokaryons where both parental cells exhibited the same type of histotypic differentiation (myogenic or erythroid) and the nuclei were in similar states of activity.

New techniques for enucleating large numbers of somatic cells can be combined with the Sendai virus fusion technique in the study of cell differentiation. Cells can be *reconstituted* by fusing new nuclei into enucleated cytoplasms or by fusing intact cells with enucleated cytoplasms. By combining nuclei and cytoplasms from cells representing different lines of cell differentiation and then analyzing phenotypic expression in the reconstituted cells, it should be possible to gain new insights into mechanisms regulating genetic expression in eukaryotic cells.

REFERENCES

1. Bolund, L., Ringertz, N. R., and Harris, H. (1969a): Changes in the cytochemical properties of erythrocyte nuclei reactivated by cell fusion. *J. Cell Sci 4*:71.
2. Bolund, L., Darzynkiewicz, Z., and Ringertz, N. R. (1969b): Growth of hen erythrocyte nuclei undergoing reactivation in heterokaryons. *Exp. Cell Res. 56*:406.
3. Bootsma, D., De Weerd-Kastelein, E. A., Veldhuisen, G., and Keyzer, W. (1972): Genetic studies on repair in Xeroderma pigmentosum cells. DNA-repair mechanisms. *Symposia Medica Heochsl.,* F. K. Schaltauer Verlag, Stuttgart.
4. Carlsson, S. A., Luger, O., Ringertz, N. R., and Savage, R. E. (1974): Phenotypic expression in chick erythrocyte × rat myoblast hybrids and in chick myoblast × rat myoblast hybrids. *Exp. Cell Res. 84*:47.
5. Carlsson, S. A., Moore, P., and Ringertz, N. R. (1973): Nucleocytoplasmic protein migration during the activation of chick erythrocyte nuclei in heterokaryons. *Exp. Cell Res. 76*:134.
6. Carlsson, S. A., Savage, R. E., and Ringertz, N. R. (1970): Behaviour of differentiated hen nuclei in the cytoplasm of rat myoblasts and myotubes. *Nature 228*:869.
7. Cook, P. R. (1970): Species specificity of an enzyme determined by an erythrocyte nucleus in an interspecific hybrid cell. *J. Cell Sci. 7*:1.
8. Darzynkiewicz, Z., and Chelmicka-Szorc, E. (1972): Unscheduled DNA synthesis in hen erythrocyte nuclei reactivated in heterokaryons. *Exp. Cell Res. 74*:131.
9. Deák, I., Sidebottom, E., and Harris, H. (1972): Further experiments on the role of the nucleolus in the expression of structural genes. *J. Cell Sci. 11*:379.
10. Ege, T., Carlsson, S. A., and Ringertz, N. R. (1971): Immune microfluorimetric analysis of the distribution of species specific nuclear antigens in HeLa-chick erythrocyte heterokaryons. *Exp. Cell Res. 69*:472.

11. Ege, T., Zeuthen, J., and Ringertz, N. R. (1973): Cell fusion with enucleated cytoplasms in chromosome identification—Techniques and applications in biology and medicine. Nobel Symposium 23. (ed. T. Caspersson and L. Zech). Academic Press, New York.
12. Ephrussi, B. (1972): *Hybridization of Somatic Cells,* Princeton Univ. Press, Princeton, New Jersey.
13. Fell, H. B., and Hughes, A. F. (1949): Mitosis in the mouse: a study of living and fixed cells in tissue culture. *Quant. J. Microscop. Sci. 90*:355.
14. Goldstein, L., and Trescott, O. H. (1970): Characterization of RNAs that do and do not migrate between cytoplasm and nucleus. *Proc. Nat. Acad. Sci. USA 67*:1367.
15. Gordon, S., and Cohn, Z. (1970): Macrophage-malanocyte heterokaryons. I. Preparation and properties. *J. Exp. Med. 131*:981.
16. Gordon, S., and Cohn, Z. (1971a): Macrophage-melanocyte heterokaryons. II. The activation of macrophage DNA synthesis. Studies with inhibitors of RNA synthesis. *J. Exp. Med. 133*:321.
17. Gordon, S., and Cohn, Z. (1971b): Macrophage-melanoma cell heterokaryons. III. The activation of macrophage DNA synthesis. Studies with inhibitors of protein synthesis and with synchronized melanoma cells. *J. Exp. Med. 134*:935.
18. Gordon, S., and Cohn, Z. (1971c): Macrophage-melanoma cell heterokaryons. IV. Unmasking the macrophage-specific receptor. *J. Exp. Med. 134*:947.
19. Harris, H. (1965): Behaviour of differentiated nuclei in heterokaryons of animal cells from different species. *Nature 206*:583.
20. Harris, H. (1967): The reactivation of the red cell nucleus. *J. Cell Sci. 2*:23.
21. Harris, H. (1970): *Cell fusion: The Dunham Lectures,* Clarendon Press, Oxford.
22. Harris, H., and Cook, P. R. (1969): Synthesis of an enzyme determined by an erythrocyte nucleus in a hybrid cell. *J. Cell Sci. 5*:121.
23. Harris, H., Sidebottom, E., Grace, D. M., and Bramwell, M. E. (1969): The expression of genetic information: a study with hybrid animal cells. *J. Cell Sci. 4*:499.
24. Harris, H., and Watkins, J. F. (1965): Hybrid cells derived from mouse and man. Artificial heterokaryons of mammalian cells from different species. *Nature 205*:640.
25. Harris, H., Watkins, J. F., Campbell, J. L. M., Evans, E. P., and Ford, C. E. (1965): Mitosis in hybrid cells derived from mouse and man. *Nature 207*:606.
26. Harris, H., Watkins, J. F., Ford, C. E., and Schoefl, G. I. (1966): Artificial heterokaryons of animal cells from different species. *J. Cell Sci. 1*:1.
27. Ikeuchi, T., and Sandberg, A. A. (1970): Chromosome pulverization in virus-induced heterokaryons of mammalian cells from different species. *J. Nat. Cancer Inst. 45*:951.
28. Jacobson, C. O. (1968): Reactivation of DNA synthesis in mammalian neuron nuclei after fusion with cells of an undifferentiated fibroblast line. *Exp. Cell Res. 53*:316.
29. Johnson, R. T., and Harris, H. (1969a): DNA synthesis and mitosis in fused cells. I. HeLa homokaryons. *J. Cell Sci. 5*:603.
30. Johnson, R. T., and Harris, H. (1969b): DNA synthesis and mitosis in fused cells. II. HeLa-chick erythrocyte heterokaryons. *J. Cell Sci. 5*:625.
31. Johnson, R. T., and Harris, H. (1969c): DNA synthesis and mitosis in fused cells. III. HeLa-Ehrlich heterokaryons. *J. Cell Sci. 5*:645.
32. Johnson, R. T., and Rao, P. N. (1970): Mammalian cell fusion. II. Induction of premature chromosome condensation in interphase nuclei. *Nature 226*:717.
33. Johnson, R. T., and Rao, P. N. (1971): Nucleo-cytoplasmic interactions in the achievement of nuclear synchrony in DNA synthesis and mitosis of multinucleate cells. *Biol. Rev. 46*:97.
34. Johnson, R. T., Rao, P. N., and Hughes, H. D. (1970): Mammalian cell fusion. III. A HeLa cell inducer of premature chromosome condensation active in cells from a variety of animal species. *J. Cell Physiol. 76*:151.
35. Kato, H., and Sandberg, A. A. (1967): Chromosome pulverization in human binucleate cells following colcemid treatment. *J. Cell Biol. 34*:35.
36. Marshall Graves, J. A. (1972a): DNA synthesis in heterokaryons formed by fusion of mammalian cells from different species. *Exp. Cell Res. 72*:393.
37. Marshall Graves, J. A. (1972b): Cell cycles and chromosome replication patterns in interspecific somatic hybrids. *Exp. Cell Res. 73*:81.
38. Matsui, S., Weinfeld, H., and Sandberg, A. A. (1971): Dependence of chromosome pulverization in virus-fused cells on events in the G_2 period. *J. Nat. Cancer Inst. 47*:401.

39. Mazia, D. (1961): Mitosis and the physiology of cell division. *The Cell.* Vol. 3. (ed. J. Brachet and A. E. Mirsky). Academic Press, New York.
40. Nichols, W. W., Levan, A., Aula, P., and Norrby, E. (1965): Chromosome damage associated with the measles virus in vitro. *Hereditas 54*:101.
41. Oftebro, R., and Wolf, I. (1967): Mitosis of bi- and multi-nucleate HeLa cells. *Exp. Cell Res. 48*:39.
42. Prescott, D. M., Myerson, D., and Wallace, J. (1972): Enucleation of mammalian cells with cytochalasin B[1]. *Exp. Cell Res. 71*:480.
43. Ringertz, N. R. (1969): Cytochemical properties of nuclear proteins and deoxyribonucleoprotein complexes in relation to nuclear function. *Handbook of Molecular Cytology,* p. 656, (ed. A. Lima-de-Faria), North Holland, Amsterdam.
44. Ringertz, N. R., and Bolund, L. (1969): "Activation" of hen erythrocyte deoxyribonucleoprotein. *Exp. Cell Res. 55*:205.
45. Ringertz, N. R., and Bolund, L. (1974): Reactivation of chick erythrocyte nuclei by somatic cell hybridization. *Int. Rev. Exp. Pathol.,* Vol. XIII (ed. G. W. Richter and M. A. Epstein). Academic Press, New York.
46. Ringertz, N. R., Carlsson, S. A., Ege, T., and Bolund, L. (1971): Appearance of human and chick nuclear antigens in chick erythrocyte nuclei during reactivation in heterokaryons with HeLa cells. *Proc. Nat. Acad. Sci. USA 68*:3228.
47. Rosenberg, M. (1972): Fusion of frog and tadpole erythrocytes. *Nature 239*:520.
48. Schwartz, A. G., Cook, P. R., and Harris, H. (1971): Correction of a genetic defect in a mammalian cell. *Nature New Biol. 230*:5.
49. Sidebottom, E., and Harris, H. (1969): The role of the nucleolus in the transfer of RNA from nucleus to cytoplasm. *J. Cell Sci. 5*:351.
50. Stadler, J. K., and Adelberg, E. A. (1972): Cell cycle changes and the ability of cells to undergo virus-induced fusion. *Proc. Nat. Acad. Sci. USA 69*:1929.
51. Stenman, S., and Saksela, E. (1969): Susceptibility of human chromosomes to pulverization induced by myxoviruses. *Hereditas 62*:323.
52. Steplewski, Z., and Koprowski, H. (1969): Development of SV_{40} coat protein antigen in nonpermissive nuclei in heterokaryocytes. *Exp. Cell Res. 57*:433.
53. Thompson, E. B., and Gelehrter, T. D. (1971): Expression of tyrosine aminotransferase activity in somatic cell heterokaryons: Evidence for negative control of enzyme expression. *Proc. Nat. Acad. Sci. USA 68*:2589.
54. Watkins, J. F., and Grace, D. M. (1967): Studies on the surface antigens of interspecific mammalian cell heterokaryons. *J. Cell Sci. 2*:193.
55. Westerveld, A., and Freeke, M. A. (1971): Cell cycle of multinucleate cells after cell fusion. *Exp. Cell Res. 65*:140.
56. Wright, W. E., and Hayflick, L. (1972): Formation of anucleolate and multinucleate cells in normal and SV_{40} transformed WI-38 by cytochalasin B. *Exp. Cell Res. 74*:187.
57. Yamanaka, T., and Okada, Y. (1966): Cultivation of fused cells resulting from treatment of cells with HVJ. I. Synchronization of the stages of DNA synthesis of nuclei involved in fused multinucleated cells. *Biken J. 9*:159.

DISCUSSION

Weiss: How frequent are the heterokaryons that produce chicken myosin?

Ringertz: They are extremely rare. I cannot give an exact frequency, but we must have examined millions of heterokaryons without finding any chick-myosin synthesis. Then, we found about 20 such cells in experiments with low calcium in the last 2 years.

Davidson: Harris had reported at one time that the nuclei of heterokaryons could be separated after fusion. If nuclei are separated after reactivation has occurred, is the erythrocyte nucleus capable of continuing DNA and RNA synthesis in the absence of the nucleus from the other parent?

Ringertz: The nuclei that Harris isolates using detergents probably would not be capable of much activity.

Davidson: I think it was claimed at one time that the nuclei could be separated physically.

Ringertz: It is possible using sucrose gradients.

Davidson: After you separate the nuclei under conditions in which they should remain active, do they remain active? In other words, if you take a heterokaryon in which the erythrocyte nuclei have been reactivated and separate out the erythrocyte nuclei, are these nuclei then capable of continued activity or does their activity depend at all times on the presence of the nucleus of a previously active cell?

Ringertz: I can only answer this question in this way: If you enucleate a HeLa cell and fuse an erythrocyte nucleus with the HeLa cell cytoplasm, the erythrocyte nucleus is capable of reactivation.

Davidson: To what extent?

Ringertz: They make RNA and, in a very few cases, also DNA. They also form a nucleolus and express chick nucleolar antigens. However, we have not found that the reactivation of the erythrocyte nucleus results in the rescue of the cytoplasm or prevents the gradual decay in protein synthesis. Such rescue may occur in experiments in which we have fused mouse lymphocytes with TK-deficient mouse cells that have been enucleated.

Attardi: Do the reactivated erythrocyte nuclei produce ribosomal RNA?

Ringertz: Yes, I think so. Harris has shown this by gradient centrifugation techniques [Harris, H., Sidebottom, E., Grace, D., and Bramwell, M. (1969): *J. Cell Sci. 4*:499]. The results suggest that there is ribosomal RNA in the erythrocyte nuclei that have been removed from heterokaryons.

Attardi: I would like to raise a general question concerning heterokaryons between chicken erythrocytes and mammalian cells. It is generally assumed that normal adult erythrocytes from birds do not produce RNA and DNA. This is true of the great majority of the cells; however, there is a proportion of the cells that produces RNA and DNA at very low rates. Is it possible that the cells that are actually involved in the heterokaryon formation are just those cells that have a very low degree of activity and that what is called reactivation is really only the acceleration of a process that was already going on? This could involve completely different mechanisms.

Ringertz: The erythrocyte nuclei found in heterokaryons immediately after fusion do not make RNA. It takes some swelling up before RNA synthesis begins.

Attardi: However, since autoradiography is used to detect nucleic acid synthesis, there is a problem of sensitivity. If you use instead large masses of cells and high specific-activity precursors, you can detect nucleic-acid synthesis in the mature erythrocyte population by biochemical means. This level of synthesis is very low and I doubt that it would be detectable by autoradiography.

Ringertz: You are right about the sensitivity of autoradiography. There is another point. If you look at immature cells representing different stages of erythropoiesis, the nucleus is larger during the early stages of erythropoiesis and then condenses to a very small size. The erythrocyte nuclei that we find in the heterokaryons are just as small and inactive at the beginning as those in the majority of erythrocytes.

Attardi: I asked whether there is any ribosomal RNA synthesized because that would be a sign of reactivation. The immature erythrocytes do not produce ribosomal RNA, but they do produce heterogeneous nuclear RNA. The presence of ribosomal RNA would therefore be a sign of reactivation.

Ringertz: We have not looked at this ourselves. However in Harris' experiments (mentioned above) there was ribsomal-RNA synthesis by the chick erythrocyte nucleus.

Coon: If I understood correctly your answer to Dr. Davidson, the cells resulting from the fusion of chick erythrocytes with enucleated cytoplasm do not divide. How long do they live?

Ringertz: Three days at the maximum.

Coon: Just as though the cytoplam had not received a nucleus?

Ringertz: We have not seen that it makes much difference to the survival of the cytoplam whether or not it has received an erythrocyte nucleus.

Siniscalco: You concluded that the results of the reconstitution experiments do not definitively prove that the nucleus is extraneous to reactivation. I would like to know Dr. Koprowski's opinion of your interpretation and your opinion of Dr. Koprowski's interpretation of his virus rescue experiments, done in exactly the same way. It seems to me that if your conclusion is correct, then one should come to the same conclusion also about the virus rescue, i.e., that this type of experiment cannot really exclude the possibility that the nucleus is actually participating in virus rescue since something produced by the nucleus and essential for virus rescue could remain after enucleation.

Koprowski: I don't think that one set of experiments has anything to do with the other. In our system either enucleation or irradiation of SV40-permissive African green monkey cells does not prevent synthesis of infectious SV40 in the cytoplasm of these cells after fusion with SV40-nonpermissive but transformed mouse cells [Croce, C., and Koprowski, H., (1973): *Virology 51*:227; K. Huebner, *in preparation*]. It is always possible that "something essential" remains in the cytoplasm after enucleation or irradiation of the cells. In the study of the virus rescue phenomenon the important thing is the treatment of the donor, nonpermissive cell. The permissive cell may be a "cytoplasmic broth" that "amplifies" the virus ready to be rescued by the treatment of the nonpermissive transformed cell. (K. Huebner, *in preparation*).

Evans: What happens if you remove the nuclei from HeLa cells and then put these back into the enucleated cytoplasms? Are the reconstituted cells still viable?

Ringertz: I don't know yet. We are trying various other combinations in which the nucleus is not inactive as is the erythrocyte nucleus. We are trying it in such a way that we can use selective media and genetic markers, so we don't want to use HeLa cells. We have had to make our own mutants for this purpose, and we do not yet have the results.

Evans: I thought Carter in fact had put some nuclei back and still failed to get cells to survive.

Ringertz: Not that I know of.

Thompson: Has anybody succeeded in putting active nuclei into cytoplasm from other cells?

Ringertz: Not that I know of.

Thompson: Have experiments been done with nuclei isolated by cytochalasin B treatment? Such nuclei are surrounded by a much reduced amount of cytoplasm.

Ringertz: If you centrifuge cells in a serum albumin gradient [Ege, T., Appels, R., and Ringertz, N. *in preparation*] containing various medium components and cytochalasin, the cells are enucleated, and the nuclei can be separated because they are denser and will sediment faster than the cytoplasms. The sedimented nuclei will have a plasma membrane and a thin rim of cytoplasm. We therefore prefer to call these structures "minicells."

Both our group and Prescott's group have been analyzing these minicells. Vital stains do not penetrate into the nuclei. They are capable of limited RNA synthesis, but they are not able to survive or to regenerate the cytoplasm by themselves. The minicells have the receptors for Sendai virus on their plasma membrane, and they can be fused with monolayers of enucleated cells.

Silagi: There was a recent report from Tatum's laboratory on the fusion of nuclei and cytoplasm into single cells using microsurgical techniques [Diacumakos, E., and Tatum, E. (1972): *Proc. Nat. Acad. Sci. 69*:2959]. Do these cells survive after fusion?

Ruddle: What they have done there is to insert cytoplasmic projections of one cell into an adjacent cell. The membranes fuse, leading to the formation of a heterokaryon or dikaryon. They did not transfer nuclei from one cell to another.

Ringertz: There is a series of interesting "reconstitution" experiments that can be done with minicells and enucleated cytoplasms, but we don't know yet how viable the reconstituted cells would be. However, even with existing techniques, there are many interesting experiments that can be done. If you have a normal cell with a small amount of cytoplasm, you can attach to it a big piece of cytoplasm from another cell. For example, the nucleated cell could be a hepatoma cell that is producing TAT, and the cytoplasm could be from a large tetraploid cell that is undifferentiated. It would be very interesting for instance to repeat the experiments of Thompson and Gelehrter [Thompson, E. B., and Gelehrter, T. (1971): *Proc. Nat. Acad. Sci. 68*:2589] to see if TAT inducibility was extinguished. We are trying to do this at present but we do not yet have any results.

Littlefield: Could you repeat what you said about the fusion of frog and tadpole cells that make hemoglobin?

Ringertz: Both cells continue to make hemoglobin after fusion, but there is a stimulation of one kind of hemoglobin and a reduction of the other.

Littlefield: Can you fuse two cells, one of which is making hemoglobin and the other of which is not?

Ringertz: Harris has done such experiments and observed that hemoglobin synthesis was switched off.

Littlefield: Have such experiments been performed involving the fusion of frog and tadpole cells?

Ringertz: Not as far as I know.

Koprowski: I am not sure it is feasible to replace nuclei of enucleated mammalian cells by other nuclei. Enucleation involves a drastic treatment with a chemical compound and shearing of the cells by centrifugation. To my knowledge no one has produced hybrids by introduction of a nucleus into the cytoplasm of an enucleated mammalian cell.

Ringertz: With chick erythrocyte nuclei, it may work. Chick erythrocyte nuclei were fused with enucleated mouse fibroblasts, and the erythrocyte nuclei were reactivated.

Koprowski: But what happens 2 weeks later?

Ringertz: The heterokaryon dies. However that may have been related to the fact that reactivation of the chick erythrocyte nucleus takes time.

Ephrussi: You referred to the fusion of enucleated mouse fibroblasts with chick erythrocyte nuclei. In previous reports, Harris seemed to imply that full reactivation, i.e., swelling of the nucleus, formation of the nucleolus and synthesis of chick antigens, does not occur unless the fibroblasts are irradiated. Why does complete reactivation occur without irradiating the fibroblasts in your experiments?

Ringertz: I don't know why. Our timetable is different from that of Harris'. We find nucleoli earlier than he does (we find them already on the second day, visible with the light microscope) and we also see chick antigens earlier than he does. It may have to do with the fact that Harris does his fusions in suspension whereas we avoid trypsinization and add erythrocytes and virus to a monolayer of cells attached to a plastic dish.

Somatic Cell Hybridization, edited by
Richard L. Davidson and Felix de la Cruz.
Raven Press, New York © 1974.

Cell Fusion in the Study of Genetic Heterogeneity of Xeroderma Pigmentosum

D. Bootsma

Department of Cell Biology and Genetics, Erasmus University, Rotterdam, The Netherlands

I will present some data that demonstrate the applicability of cell fusion in the study of genetic heterogeneity in genetic diseases in man. We have fused cells originating from different patients suffering from the disease, xeroderma pigmentosum, which manifests itself in cell culture as the inability to repair ultraviolet light-induced DNA damage. Our interest is in the mechanism of DNA repair in mammalian cells following ultraviolet or X-irradiation, and we have been looking for different mutations affecting one or different steps in DNA repair.

Two different techniques were applied to demonstrate DNA repair. By use of ^{3}H-thymidine labeling and autoradiography, repair DNA synthesis can be visualized following UV-irradiation and the number of grains over the cells is a measure of the repair activity. In the second technique, DNA synthesized semiconservatively in the presence of bromodeoxyuridine can be separated by density gradient centrifugation from DNA synthesized as a result of repair when synthesis involves small stretches of both DNA strands.

It was shown by Cleaver (2) that cells from patients with xeroderma pigmentosum show a much lower level of repair DNA synthesis than normal cells. The skin of these patients is very sensitive to sunlight resulting in skin cancers, changes in pigmentation, and other characteristic lesions. We considered the possibility that different mutations affecting the same or different enzymes involved in DNA repair could all lead to the symptoms of xeroderma pigmentosum. This possibility could be tested by cell fusion experiments. If different mutations were involved, two mutant genotypes might be able to cooperate with each when existing in a common cytoplasm to express a wild type phenotype. These experiments were carried out in our laboratory by De Weerd-Kastelein, Kleijer, and Keijzer; a part of this work has recently been published (3, 4).

To avoid the problems related to the isolation of hybrid cell lines, binucleate heterokaryons were studied that are present in relatively high frequency during the first few days following cell fusion. In order to identify the nature of the binucleated cells, we fused together cells from xeroderma pigmentosum patients of opposite sexes. In these combinations, the male and female nuclei could be distinguished on the basis of the quinacrine staining of the sex chromosomes during interphase.

We first fused cells from a classic xeroderma (female) patient, who only shows the skin lesions, with cells from a (male) patient having the De Sanctis-Cacchione type of xeroderma pigmentosum. In the latter form of XP severe neurological abnormalities are present in addition to the skin lesions. Three types of binucleated cells might be expected from this fusion: homokaryons of the female parent strain that would show no DNA repair after UV exposure; homokaryons of the male cell strain that would also have no DNA repair; and heterokaryons with both a female and a male nucleus that would show DNA repair if complementation were to occur. When cells from two patients with the classic form of xeroderma pigmentosum were fused, or when cells from two patients with the De Sanctis-Cacchione form of the disease were fused, no heterokaryons were observed that showed tritiated thymidine labeling following UV-exposure (Table 1). Labeling was only observed when cells from a classic patient were fused with cells from a De Sanctis-Cacchione patient (3). Similar results

TABLE 1. *Percentages of fused binucleated cells with weakly labeled nuclei after UV-exposure (both nuclei > 8 grains)*

Fused populations		0 erg/mm²	100 erg/mm²
Classic XP/Classic XP:	XP4RO/XP9RO	0	8
	XP4RO/XP16RO	0	2
	XP4RO/XP20RO	0	4
	XP4RO/XP21RO	0	4
	XP9RO/XP16RO	0	4
	XP9RO/XP20RO	0	0
	XP16RO/XP20RO	2	0
DSC/DSC	XP12RO/XP4LO	0	0
	XP12RO/XPPKSF	0	0
	XP4LO/XP17SF	0	0
	XP4LO/XPPKSF	0	0
	XP17SF/XP25RO	0	0
	XP25RO/XP26RO	0	0
	XP26RO/XPPKSF	0	0
Classic XP/DSC	XP4RO/XP12RO	0	36
	XP4RO/XP4LO	0	38
	XP4RO/XP25RO	0	36
	XP9RO/XP12RO	0	30
	XP16RO/XP12RO	0	36
	XP16RO/XP4LO	0	36
	XP20RO/XP12RO	0	44

Cells were irradiated 2 days after fusion with 100 erg/mm² UV light and incubated for 2 hr in ^{3}H-thymidine containing medium. In the autoradiographic preparations only cells with 0 to 50 grains over each of both nuclei were counted; binucleated cells with 8 to 50 grains over each of both nuclei were counted as labeled.

were obtained in seven different fusions using cells from unrelated patients (four classic XP and three De Sanctis-Cacchione XP, Table 1).

It is not possible to study repair at a single cell level with the density gradient technique, so the fusion procedure had to be modified. By using a higher multiplicity of virus and a higher concentration of cells during the fusion, heterokaryons occur at such a frequency that about 60% of the nuclei can carry out repair DNA synthesis in fusions between classic and De Sanctis-Cacchione cells. The total cell population can then be labeled with BUDR and the DNA separated by density gradient centrifugation (4).

Using this technique, it was again shown that repair replication occurs only when cells from a classic xeroderma patient (XP4RO) are fused with cells from a De Sanctis-Cacchione patient (XP25RO). Fusions between cells of different classic patients or between cells of different De Sanctis-Cacchione patients produced only cells incapable of repair synthesis (Fig. 1).

We recently obtained evidence for a third complementation group by studying cells from a xeroderma patient with an intermediate level of DNA repair, our XP2RO strain (1). These cells have been fused with cells of a classic xeroderma patient and the cells from a De Sanctis-Cacchione patient. In the heterokaryons the XP2RO nuclei show the same degree of labeling as do normal control cells, and the nuclei of the other xeroderma type in the heterokaryons also performed normal levels of repair synthesis (Table 2).

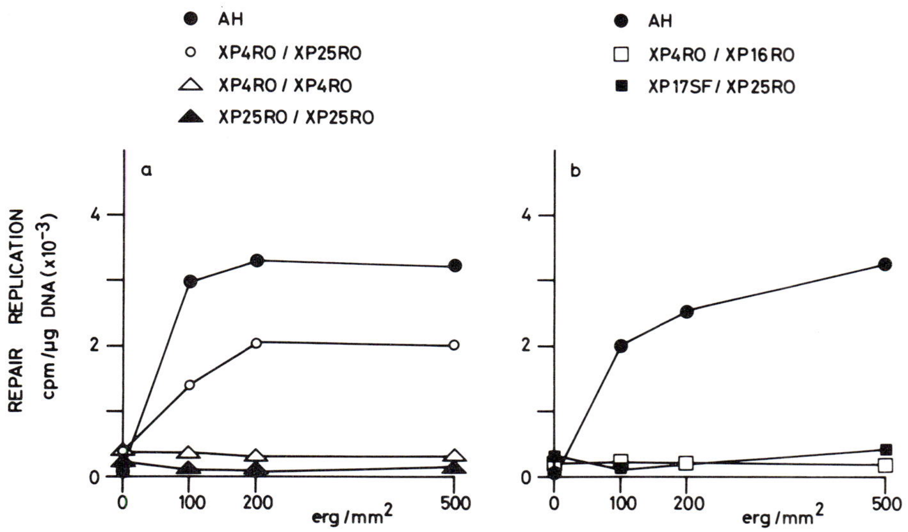

FIG. 1. Repair replication, measured in NaI isopycnic density gradients expressed as specific radioactivity (cpm/μg DNA), in normal human cells (AH) and in cell population obtained after fusion of different xeroderma pigmentosum cells. XP4RO and XP16RO are derived from patients with classic XP and XP17SF and XP25RO originate from De Sanctis-Cacchione patients. (From ref. 4.)

TABLE 2. *Grain counts per nucleus in mononucleated control and parental cell strains after exposure to 100 erg/mm² UV-light*

Cell strain	Grain counts (mean ± SEM)	Number of nuclei
Control	26.1 ± 0.7	$n = 100$
XP2RO (XX)	13.7 ± 0.6	$n = 100$
XP16RO (XY)	3.9 ± 0.2	$n = 100$

Grain counts per nucleus in atebrin diagnosed binucleated cells after exposure to 100 erg/mm² UV-light

Fusion	Diagnosed nucleus	Grain counts (mean ± SEM)	Number of nuclei
XP2RO/XP16RO	XX in XX/XX	15.0 ± 0.7	$n = 96$
	XY in XY/XY	4.2 ± 0.3	$n = 88$
	XX in XX/XY	26.4 ± 0.8	$n = 94$
	XY in XX/XY	28.0 ± 0.8	$n = 94$

The cells were exposed to 100 erg/mm² UV-light, in fused cell populations 2 days after fusion; ^{3}H-thymidine labeling was carried out 1 hr before and 2 hr after irradiation. XP2RO cells originate from a moderately severe classic case of XP. The average number of grains per nucleus and the standard error of the mean (SEM) in parental mononucleated cells was estimated by including only cells with 50 or less grains over their nuclei. In the fused binucleated cells this estimation was performed by counting only cells with 0 to 50 grains over each of both nuclei.

On the basis of these studies, three complementation groups have been defined in xeroderma pigmentosum. It is not known whether these mutations are located in the same or in different genes, affecting the same or different enzymes involved in DNA repair. Further biochemical and genetical studies are required to solve that problem.

REFERENCES

1. Bootsma, D., Mulder, M. P., Pot, F., and Cohen, J. A. (1970): Different inherited levels of DNA repair replication in xeroderma pigmentosum cell strains after exposure to ultraviolet irradiation. *Mutation Res. 9*:507–516.
2. Cleaver, J. E. (1968): Defective repair replication of DNA in xeroderma pigmentosum. *Nature 218*:652–656.
3. De Weerd-Kastelein, E. A., Keijzer, W., and Bootsma, D. (1972): Genetic heterogeneity of xeroderma pigmentosum demonstrated by somatic cell hybridization. *Nature 238*:80–83.
4. De Weerd-Kastelein, E. A., Kleijer, W. J., Sluyter, M. L., and Keijzer, W. (1973): Repair replication in heterokaryons derived from different repair deficient xeroderma pigmentosum strains. *Mutation Res. 19*:237–243.

DISCUSSION

Coon: Jay Robbins and Kenneth Kraemer of the Dermatology Branch at NCI and I were inspired by Bootsma's observations to look at a series of patients with xeroderma pigmentosum. We have evidence for four complementation groups in a series of 11 patients. Patients with neurological abnormalities fall into three of these complementation groups. Each of the four groups is characterized by a different rate of DNA repair: less than 2% of the normal repair rate (Group A); 3 to 7% (Group B); 15 to 25% (Group C); and 25 to 55% (Group D).

Bootsma: Did you find the normal level of repair in all heterokaryons containing nuclei of different groups?

Coon: The data are not complete, but the repair in such heterokaryons approaches the normal level. It is always higher than the highest level in the group D patients. We cannot yet rule out the possibility that there may be incomplete complementation in certain combinations.

Basilico: What is actually the defect in xeroderma pigmentosum?

Bootsma: In both the classic form and the De Sanctis-Cacchione form, there is good evidence that thymine dimers are not excised from the DNA. In recent experiments, M. C. Paterson, P. H. M. Lohman and M. L. Sluyter at Rijswijk, the Netherlands (Mutation Res., *in press*) have followed the removal of thymine dimers from the DNA by isolating the DNA at different time intervals after UV-exposure and treating it with extracts from Micrococcus luteus with UV-specific endonuclease activity. Where thymine dimers are present, the DNA is broken, which can be measured on alkaline sucrose gradients. They have shown that neither the classic nor the De Sanctis-Cacchione type are able to excise thymine dimers. The dimers were present for at least 24 hr after irradiation.

Gelehrter: What is the level of unscheduled DNA synthesis in heterozygotes? Also is there heterogeneity in XP-2 patients or is that patient completely unique?

Bootsma: We have only one case in which there is a decreased level of unscheduled DNA synthesis in heterozygotes. That was found in both parents of our De Sanctis-Cacchione patients. In five or six different heterozygotes, we found exactly the same level of repair. Clinically, there is a difference between the XP-2 patient and the other patients. This patient showed the first symptoms of the disease at a much later age than the other patients, suggesting that she was less severely ill than the others.

Minna: Is there any evidence in other clinical conditions, for example acquired severe photosensitivity, that there is a defect in DNA repair?

Bootsma: Not as far as I know. I think that nearly all the diseases known to involve UV sensitivity of the skin have been studied now, and there is no indication in any of them of defective DNA repair. It must be remembered, however, that we have only a very limited number of techniques to study DNA repair. There could be defects in other steps of DNA that we cannot yet measure.

Somatic Cell Hybridization, edited by
Richard L. Davidson and Felix de la Cruz.
Raven Press, New York © 1974.

The Expression of Macrophage Specific Functions in Fused Cells

Saimon Gordon

Rockefeller University, New York, New York 10021

I will discuss studies of macrophage specific functions in heterokaryons. The macrophage arises from bone marrow stem cells, undergoes several divisions, and passes into the bloodstream as an immature monocyte (5). The monocytes are then widely distributed among the body tissues, including the peritoneal cavity where the cells exist in a mature resting state. The process of macrophage differentiation can occur in liquid culture systems of bone marrow cells in the presence of special colony-stimulating factors.

Mature macrophages are most easily obtained by washing out the peritoneal cavity of the mouse without making an inflammatory exudate and separating the macrophages from lymphocytes by allowing the macrophages to adhere to a substrate. We obtained pure populations of primary macrophages, which are normal diploid cells, unlike most of the differentiated cells used in fusion experiments. These cells form a stable nondividing population, which is arrested in G_0. The cells synthesize no DNA over a period of many weeks but continue to synthesize RNA and protein and to express several differentiated functions.

A special receptor is present on the plasma membrane of the macrophage that recognizes uniquely the Fc region of the antibody molecule, especially in certain classes of IgG immunoglobulins. For example, sheep red blood cells coated with a specific 7S antibody will attach to this receptor. This attachment triggers a phagocytic response which is only seen in specialized phagocytic cells like macrophages and polymorphonuclear leukocytes. The receptor is very stable and is exhibited by all the cells in a culture. Another useful marker is the plasma membrane ATPase.

We were interested in seeing the results of fusing nondividing, differentiated macrophages that are capable of maintaining the differentiated state *in vitro,* with multiplying, undifferentiated cells. We began by making heterokaryons between mouse macrophages and nonpigmented, mouse melanoma cells (1). Within 3 hr after fusion, DNA synthesis is initiated in the macrophage nucleus. The activation of the macrophage nucleus requires RNA synthesis by the melanoma nucleus and the entry of proteins from the melanoma cytoplasm into the macrophage nucleus (2, 3). We have also looked at the membrane ATPase. A few hours after fusion, the ATPase activity is distributed in patches. However, about 10 to 12 hr after fusion, the ATPase is diffusely distributed over the cell membrane.

Our main interest was the fate of the macrophage specific phagocytic receptor in the heterokaryons. The melanoma cells do not have this receptor. Immediately after fusion between macrophages and melanoma cells, the heterokaryons have full receptor activity. With time, however, the heterokaryons progressively lose receptor activity, so that within 24 hr, the heterokaryons can no longer bind sensitized red blood cells (1). In macrophage homokaryons, on the other hand, receptor activity is retained. In homokaryons with different numbers of nuclei, as the number of macrophage nuclei increases, the cell surface area and receptor activity increase in parallel.

Figure 1 shows the decay of receptor activity in heterokaryons containing one, two, or three macrophage nuclei and one melanoma nucleus. A dosage effect, depending on the number of macrophage nuclei in the heterokaryon, is observed, but in all cases receptor activity is progressively lost.

If macrophage-melanoma heterokaryons in which receptor activity has disappeared are treated with trypsin under conditions of mild proteolysis, receptor activity reappears (Fig 2) (4). Receptor activity is then lost again

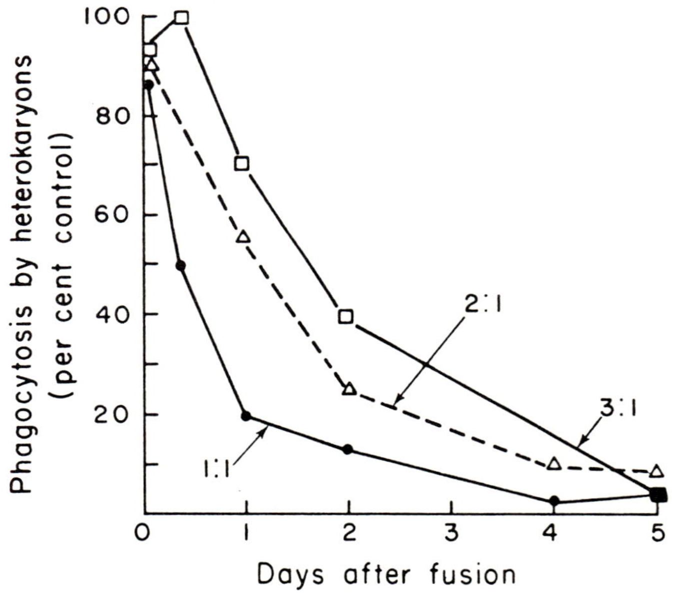

FIG. 1. Phagocytosis by macrophage-melanoma cell heterokaryons. Macrophage homokaryons containing the same number of macrophage nuclei served as controls. (From ref. 1.)

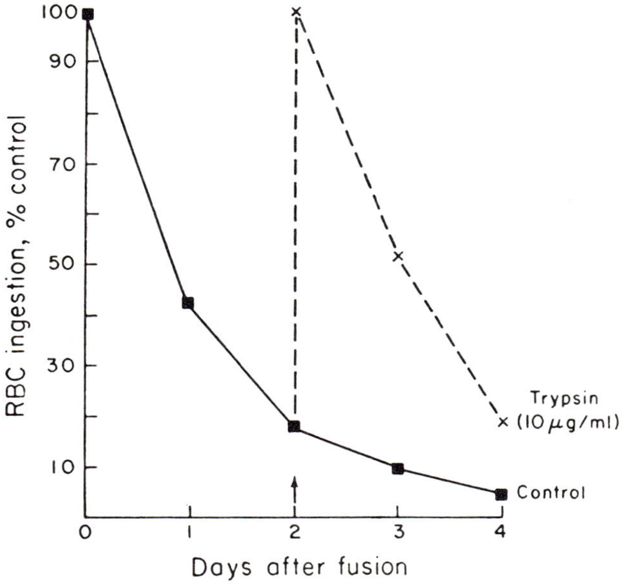

FIG. 2. Recovery of the Fc receptor in heterokaryons. Trypsin treatment 2 days after fusion (arrow). (From ref. 4.)

as the cells are maintained in culture. Treatment of the heterokaryons with neuraminidase does not unmask the receptors.

Masking of the receptors was shown not to be due to adsorption of some factor from the medium but is an active process. Masking is inhibited by cycloheximide treatment of the heterokaryons but reoccurs when the cycloheximide is removed from the medium. At this time, the receptor activity of the cells can again be activated by trypsin treatment.

In other experiments, we have demonstrated that RNA synthesis is also necessary for masking the Fc receptor in the heterokaryons. The melanoma cell nucleus is responsible for this RNA synthesis. If the melanoma cell nucleus is irradiated with ultraviolet light prior to fusion, the receptors are not masked; if the macrophage nucleus is irradiated, masking of the receptors in the heterokaryons proceeds unimpaired.

Masking of the Fc receptor is not unique to the melanoma cell. Masking also occurs in heterokaryons between macrophages and cells of a number of other lines, such as Ehrlich ascites tumor cells, human lymphoid cell lines, and BUDR-treated melanoma cells. It would appear that an active nucleus is necessary for masking. If a chick erythrocyte with an inert nucleus is fused with macrophages, masking does not occur.

These experiments indicate that some type of alteration in the cell surface properties of heterokaryons is occurring and that it may be mediated by the nucleus of a foreign cell. Unfortunately, the turnover time of the Fc receptor in these heterokaryons is unknown, and we do not know whether the receptor is resynthesized in the heterokaryons.

If irradiated chick fibroblasts are fused with macrophages, the receptor activity of the heterokaryons declines slowly over several days. The decline is slower than in heterokaryons with melanoma cells. In these heterokaryons, receptor activity is not recovered by trypsin treatment of the cells. The simplest explanation is that the decline in receptor activity in these heterokaryons represents the true turnover time of the receptor without any complications by masking. Alternative explanations are, however, possible, and problems such as incomplete synthesis could be considered.

Because of the complexity of cell-surface markers, we decided to investigate other macrophage specific functions. Fleming had noted that white blood cells produce large amounts of lysozyme, an enzyme that lyses *Micrococcus lysodeikticus.* Mass cultures of macrophages secrete large amounts of lysozyme, with up to 2% of the cell mass being released as lysozyme into the medium per day. This enzyme is a fairly specific macrophage marker. Polymorphonuclear leukocytes, which have a common stem-cell origin, also contain the enzyme, but our cultures are free of granulocytes. Lymphoid cells, fibroblasts, neuroblastoma cells, and other tissue-culture cell lines do not produce lysozyme, although other cells such as those of the chick oviduct may possibly secrete the enzyme.

Lysozyme is a single chain molecule with a molecular weight of 14,000. It is very cationic, and there are no known genetic variants except perhaps in some birds. We have antibodies to chick lysozyme that show no cross-reaction with either rodent or human lysozyme. We also have antibodies to the mouse enzyme, which differ from the human enzyme by about 25% in amino acid composition. There is some cross-reaction with the human enzyme, but the sera can be made species specific by cross-adsorption.

Lysozyme occurs both intracellularly and extracellularly. Within the cell, the enzyme is probably present in a granular form. As shown by fluorescent antibody techniques, all macrophages in culture contain lysozyme and continue to produce the enzyme. Human monocytes also continue to express lysozyme as they mature in culture.

For these reasons, we feel that lysozyme will be a useful macrophage marker in heterokaryon and hybrid cells. Finally, in future studies, if this enzyme is expressed in interspecies hybrids where one parent cell is a human white blood cell, the presence of the enzyme could distinguish hybrids of which the parental cell is a lymphoid cell from those in which the parental cell is a monocyte or granulocyte.

REFERENCES

1. Gordon, S., and Cohn, Z. (1970): Macrophage-melanocyte heterokaryons. I. Preparation and properties. *J. Exp. Med. 131*:981–1003.

2. Gordon, S., and Cohn, Z. (1971a): Macrophage-melanocyte heterokaryons. II. The activation of macrophage DNA synthesis. Studies with inhibitors of RNA synthesis. *J. Exp. Med.* *133*:321–338.
3. Gordon, S., and Cohn, Z. (1971b): Macrophage-melanoma cell heterokaryons. III. The activation of macrophage DNA synthesis. Studies with inhibitors of protein synthesis and with synchronized melanoma cells. *J. Exp. Med.* *134*:935–946.
4. Gordon, S., and Cohn, Z. (1971c): Macrophage-melanoma heterokaryons. IV. Unmasking the macrophage-specific membrane receptor. *J. Exp. Med.* *134*:947–962.
5. Gordon, S., and Cohn, Z. (1972): The macrophage. In: *Int. Rev. Cytol.*, vol. 36, Academic Press, New York.

DISCUSSION

Watkins: I would like to mention briefly a hybrid line we have made between Ehrlich ascites tumor cells and macrophages. The hybrids have about 90 chromosomes, including approximately 77 from the Ehrlich cell and 20 from the macrophage (the expected total would be about 120 chromosomes). About 40% of the hybrids will adsorb sensitized erythrocytes, so they seem to have the receptor described by Gordon. The adsorption can be blocked by rabbit IgG. If the cells are mildly treated with trypsin, the percentage of cells adsorbing the sensitized erythrocytes goes up to around 85 to 90%. Upon further cultivation, masking occurs again. As Gordon described for heterokaryons, the masking can be blocked by cycloheximide.

Thompson: Has trypsin treatment ever been carried out in the presence of cycloheximide or actinomycin or analogous drugs, i.e., has the unmasking experiment ever been done in the presence of inhibitors of macromolecular synthesis?

Gordon: No.

Wiener: You have a hybrid between macrophages and tumor cells?

Gordon: No. We have produced mouse macrophage—mouse L cell fibroblast hybrids [Gordon, S., Ripps, C., and Cohn, Z. (1971): *J. Exp. Med.* *134*:1187].

Wiener: What was the evidence that macrophages were actually involved in the fusion?

Gordon: The most useful marker was the presence of an intermediate level of ATPase in the hybrid. (The L cell parent had only about 2% of the activity of the macrophage parent.) Other macrophage genes that are not cell type specific were also expressed in that hybrid.

Wiener: Was this level maintained during the whole period of the experiments?

Gordon: Yes.

Pontecorvo: Did you say that the heterokaryon secretes lysozyme?

Gordon: No, only the macrophages. We have established this as a marker for future heterokaryon systems.

Siniscalco: Dr. Watkins, what about the hybrids between Ehrlich ascites cells and macrophages?

Watkins: We have not yet tested lysozyme, but I am hoping that Gordon will do this for us.

Somatic Cell Hybridization, edited by
Richard L. Davidson and Felix de la Cruz.
Raven Press, New York © 1974.

Fusion of Mouse Blastomeres

Christopher Graham

Department of Zoology, University of Oxford, South Park Road, Oxford OX1 3PS, England

I would like to talk about the fusion of somatic cells with mouse eggs and also about the fusion of mouse blastomeres with each other. The general aim of these experiments is to develop a technique for the replacement of the zygote nucleus with the nucleus of a particular cell type from either an adult or an embryonic animal. If such a technique were to work, then one might be able to test the ability of nuclei from immunoglobulin producing cells or from malignant cells to promote normal development.

I will describe the various processes that must be achieved in order to obtain complete nuclear transfer in a mammal. I may say in advance that these processes have not yet been put together in a complete experiment.

First of all, there is the process of cell fusion, which involves slightly different techniques from those that are used with cells in culture. The cells are treated as individuals.

The two differences from the standard cell-fusion technique are (a) the mammalian egg is covered with zona pellucida which must be removed with pronase before the start of the experiment, and (b) in order to promote fusion, the two cells are held together for a short time at 5°C before the temperature is raised to 37°C and fusion progresses. With this technique, it is not difficult to obtain nearly 100% fusion (5).

The first experiments were attempts to show that the fusion process itself did not damage the development of the normal mouse egg. These involved the fusion of pairs of blastomeres from the two-cell stage of development. In the best experiments performed at the two-cell stage, 90% of the fusion products will develop into blastocysts.

Figure 1 shows blastocysts that developed from these fusions. On the basis of microdensitometry and chromosome analysis, we have shown that these are complete tetraploids. The first conclusion from this kind of work is that early embryonic cells can be fused together and retain the ability to develop into blastocysts.

Following the fusion and the transfer of these blastocysts back to a recipient animal, the blastocysts will usually give rise to trophoblast. They will only rarely form yolk sac and embryo (Graham, *unpublished*). Thus the possibility arose that the fusion conditions, although permitting blastocyst

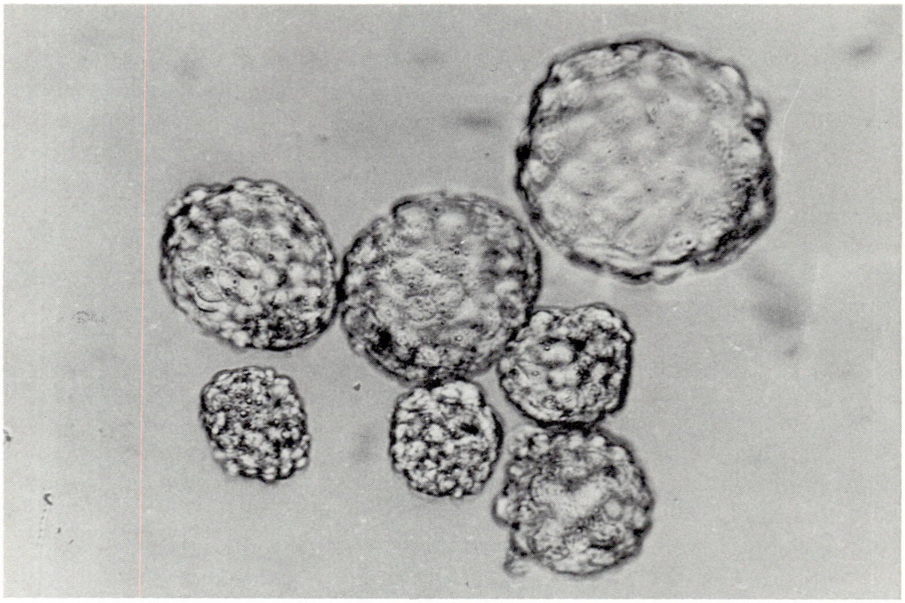

FIG. 1. Blastocysts developing from fusion of pairs of blastomeres from two cell stage.

development, in some way damaged the cells so that their subsequent potential was limited.

To find out if this really was the case, we fused together two cells at the two-cell stage from a brown mouse strain. Taking these tetraploid cells, we made chimeras with diploid cells from a white mouse strain. The diploid cells had also been exposed to the fusion conditions, but they had not fused. Six of these chimeras developed into blastocysts. These were transferred back to a recipient mouse, and from them three white mice were born. This rather limited experiment shows that the fusion conditions themselves are not the cause of restriction of development potential, but rather the restriction has something to do with the tetraploid cell itself, which was formed by the fusion of blastomeres (7). This conclusion may no longer be correct; a single tetraploid mouse has been born after the inhibition of cytokinesis at the four-cell stage by cytochalasin B (13). However many tetraploids produced by cytochalasin die after implantation and tetraploidy itself might still account for the failure of these fused blastomeres to develop to term.

The second process that I would like to discuss is the activation of the egg. The egg cytoplasm, of course, must be activated to start development in order to activate the transferred nucleus fused with it. There are several methods of activating eggs to develop parthenogenetically. One method is to stimulate the eggs with a small electric shock while they are still *in vivo,* inside the oviduct (15). Another method involves an *in vitro* stimulus. The eggs are exposed to hyaluronidase and an osmotic shock and then subsequently cultured in normal medium (4, 6–9).

Following *in vitro* activation by this method, 100% of the female pronuclei enter DNA synthesis within 4 to 5 hr after activation (Graham, *unpublished*). Up to that stage then, activation is normal. Of the activated eggs, only about 9% develop into blastocysts. It is therefore possible that there is something inadequate about the activation stimulus, with regard to promoting normal development. We do not consider this a likely possibility, because chromosome analysis has shown that many of these activated eggs have a highly abnormal chromosome complement. This is probably part of the cause of failure of development (7).

Other workers would not agree with this interpretation. For instance Mintz and Gaerhart (11) have found that the zona reaction to electrical activation is not the same as the zona reaction to fertilization. Also Tarkowski (14) has drawn attention to the possibility that the cortical reaction may not be normal. However parthenogenetic activation is equivalent to fertilization in that Conconavalin A binding sites are exposed on the cell surface after both stimuli (12).

The most advanced stage of development obtained when these activated eggs have been transferred back into the uterus, in about 10% of the cases, resembles something that an imaginative embryologist might describe as a disorganized day 7 embryo. Following electrical activation *in vivo*, Tarkowski et al. (15) have obtained a six-somite embryo.

These activation experiments have shown that, following activation by the *in vitro* method, DNA synthesis is induced in 100% of the nuclei, that 9% of the eggs develop into blastocysts, but that very few of those blastocysts develop very much further. However, it appears that we have a good technique for activating egg cytoplasm.

After a foreign nucleus has been introduced into the egg cytoplasm, the reactivation of that nucleus must occur so that it is able to divide in time with the cleavage of the egg cytoplasm. We have performed experiments in which adult cells have been fused into fertilized and unfertilized one-cell mouse eggs.

Figure 2 shows a fusion between a cell taken from the spleen and a one-cell stage egg about 11 hr after fusion. The spleen nucleus has not swollen to the same extent as the pronucleus. Reactivation is therefore slow, compared with the reactivation of the sperm nucleus. From a large number of experiments we conclude that the introduced nucleus is induced to enter DNA synthesis, that it is induced to divide in time with the resident nucleus, but that it is not reactivated (in terms of swelling) to the same extent as the normal pronucleus. Other workers have also obtained only limited swelling of the fused cell nucleus (1, 2, 10).

We have also fused adult nuclei into two-cell embryos. In this case, there

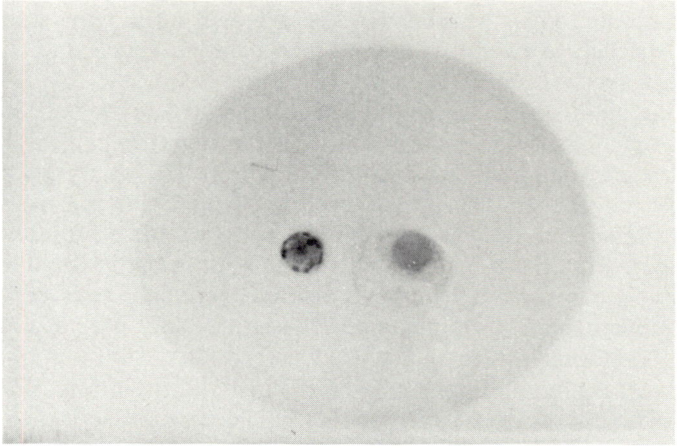

Fig. 2. Spleen cell nuclei fused with one-cell egg (activated, unfertilized) 10 hr after fusion.

is a different picture. Initially the nuclei swell. However, as each of the blastomeres at the two-cell stage enters mitosis, the introduced nuclei become pycnotic and do not divide in time with the resident nuclei. It appears that at least part of the explanation of this phenomenon involves the absence of the G_1 phase between the one-cell stage and the eight-cell stage. Therefore in this period, nuclei are introduced only into S or G_2 cytoplasm, and the nuclei are not reactivated (3).

The last process which must be achieved in order to obtain complete nuclear transfer is the enucleation of the one-cell egg. For this several techniques have been used. We have done experiments with UV irradiation of the whole egg, similar to the technique used for nuclear transfer in amphibia. Following irradiation the egg pronucleus becomes pycnotic but after the introduction of a new nucleus by fusion, we have not seen the eggs go through more than three or four mitotic divisions.

I would like to conclude by saying that the reactivation of the adult cell nucleus may be the point at which the technique of nucleus transfer with early mouse eggs becomes extremely difficult, if not impossible. However nuclear transplant experiments may be able to contribute to somatic cell genetics by permitting the fusion of adult cells with normal diploid one-cell eggs with the result that parts of the somatic cell's genome could be carried through development. This might give rather interesting possibilities in an interspecies cross.

REFERENCES

1. Baranska, W., and Koprowski, H. (1970): Fusion of unfertilized mouse eggs with somatic cells. *J. exp. Zool. 174*:1.

2. Bernstein, R. M., and Mukherjee, B. B. (1972): Control of nuclear RNA synthesis in 2-cell and 4-cell mouse embryos. *Nature 238*:457–459.
3. Graham, C. F. (1969): The fusion of embryonic and adult mouse cells. *Symposia of the Wistar Institute* 9:19–35.
4. Graham, C. F. (1970): Parthenogenetic mouse blastocysts. *Nature 226*:165–167.
5. Graham, C. F. (1971a): Virus assisted fusion of embryonic cells. *Karolinska Symposium on Research Methods in Reproductive Biology, 3,* 154–157.
6. Graham, C. F. (1971b): Experimental early parthenogenesis in mammals. *Advances in the Biosciences 6*:87–97.
7. Graham, C. F. (1972): Genetic manipulation of the early mouse embryo. *Advances in the Biosciences 8*:263–277.
8. Kaufmann, M. H. (1973a): Parthenogenesis in the mouse. *Nature, 242*:475–476.
9. Kaufmann, M. H. (1973b): Timing of the first cleavage division of haploid mouse eggs, and the duration of its component stages. *J. Cell. Sci. in press.*
10. Lin, T. P., and Florence, J., and Oh, J. O. (1973): Cell fusion induced by virus within the zona pellucida of mouse eggs. *Nature 242*:47–49.
11. Mintz, B., and Gaerhart, J. D. (1973): Subnormal zona pellucida changes in parthenogenetic mouse embryos. *Devel. Biol. 31*:178–184.
12. Pienkowski, M., and Koprowski, H. (in press): Study of the growth regulation of preimplantation mouse embryos using Concanavalin A. *J. Nat. Cancer Inst.*
13. Snow, M. H. W. (in press): Tetraploid mouse embryos produced by cytochalasin B during cleavage. *Nature.*
14. Tarkowski, A. K. (1971): Recent studies on parthenogenesis in the mouse. *J. Reprod. Fert. Suppl. 14*:31–39.
15. Tarkowski, A. K., Witkowska, A., and Nowicka, J. (1970): Experimental parthenogenesis in the mouse. *Nature 226*:162–165.

DISCUSSION

Littlefield: Could you say something about the possibility of starting haploid cell lines from your parthenogenetic embryos?

Graham: We have attempted to derive teratocarcinomas from haploid blastocysts. The problem was that it was extremely difficult to get progressive teratocarcinomas even from diploid blastocysts in the particular strain we were using. At present, we are using strains that routinely give rise to progressive teratocarcinomas from diploid blastocysts and are attempting to induce teratocarcinomas from haploid blastocysts.

AUTHOR INDEX*

* Italic numbers: contributions of participants. Other numbers: complete journal references.

SUBJECT INDEX

H

U

V

X-Z